DYNAMICS AND STRUCTURE
OF QUIESCENT SOLAR PROMINENCES

ASTROPHYSICS AND SPACE SCIENCE LIBRARY

A SERIES OF BOOKS ON THE RECENT DEVELOPMENTS
OF SPACE SCIENCE AND OF GENERAL GEOPHYSICS AND ASTROPHYSICS
PUBLISHED IN CONNECTION WITH THE JOURNAL
SPACE SCIENCE REVIEWS

Editorial Board

R.L.F. BOYD, *University College, London, England*

W. B. BURTON, *Sterrewacht, Leiden, The Netherlands*

C. DE JAGER, *University of Utrecht, The Netherlands*

J. KLECZEK, *Czechoslovak Academy of Sciences, Ondřejov, Czechoslovakia*

Z. KOPAL, *University of Manchester, England*

R. LÜST, *European Space Agency, Paris, France*

L. I. SEDOV, *Academy of Sciences of the U.S.S.R., Moscow, U.S.S.R.*

Z. ŠVESTKA, *Laboratory for Space Research, Utrecht, The Netherlands*

VOLUME 150
CURRENT RESEARCH

DYNAMICS AND STRUCTURE OF QUIESCENT SOLAR PROMINENCES

Edited by

E. R. PRIEST

Mathematical Sciences Department, St. Andrews, Scotland

KLUWER ACADEMIC PUBLISHERS

DORDRECHT / BOSTON / LONDON

Library of Congress Cataloging in Publication Data

```
Dynamics and structure of quiescent solar prominences / edited by E.R.
  Priest.
      p.    cm. -- (Astrophysics and space science library)
    Bibliography: p.
    Includes index.
    ISBN 902772833X.   ISBN 9027728348 (pbk.)
    1. Sun--Prominences--Congresses.   I. Priest, E. R. (Eric Ronald),
  1943-    .  II. Series.
  QB526.P7D96  1988
  523.7'5--dc19                                              88-25220
                                                                  CIP
```

ISBN 90-277-2833-X

Published by Kluwer Academic Publishers,
P.O. Box 17, 3300 AA Dordrecht, The Netherlands.

Kluwer Academic Publishers incorporates
the publishing programmes of
D. Reidel, Martinus Nijhoff, Dr W. Junk and MTP Press.

Sold and distributed in the U.S.A. and Canada
by Kluwer Academic Publishers,
101 Philip Drive, Norwell, MA 02061, U.S.A.

In all other countries, sold and distributed
by Kluwer Academic Publishers Group,
P.O. Box 322, 3300 AH Dordrecht, The Netherlands.

All Rights Reserved
© 1989 by Kluwer Academic Publishers
No part of the material protected by this copyright notice may be reproduced or
utilized in any form or by any means, electronic or mechanical
including photocopying, recording or by any information storage and
retrieval system, without written permission from the copyright owner.

CONTENTS

PREFACE ix

ACKNOWLEDGEMENTS x

CHAPTER 1   INTRODUCTION TO QUIESCENT SOLAR PROMINENCES (E R Priest)   1

1.1   Basic Description   1
   1.1.1 Different Types   1
   1.1.2 Properties   3
   1.1.3 Development   3
   1.1.4 Structure   4
   1.1.5 Eruption   4
1.2   Basic Equations of MHD   7
   1.2.1 Magnetohydrostatics   10
   1.2.2 Waves   11
   1.2.3 Instabilities   12
1.3   Prominence Puzzles   12

CHAPTER 2   OVERALL PROPERTIES AND STEADY FLOWS (B Schmieder)   15

2.1   Basic Properties   15
   2.1.1 Description and Classification   15
   2.1.2 Fine Structure in Hα 17
   2.1.3 Evolution of Filaments During the Solar Cycle   19
2.2   Physical Characteristics : Density and Temperature   22
   2.2.1 Density and Ionization Degree   22
   2.2.2 Non LTE Models   24
   2.2.3 Turbulent Velocity and Electron Temperature   26
2.3   Velocity Field and Mass Flux   28
   2.3.1 Instrumentation   28
   2.3.2 Hα Profile Analysis   31
   2.3.3 Vertical Motions   31
   2.3.4 Horizontal Motions   34
   2.3.5 Oscillations   37
2.4   Instability   38
   2.4.1 Disparition Brusque of Filaments   38
   2.4.2 Model Support   45
   2.4.3 Post-Flare Loops and Loop Prominences   45
2.5   Conclusion   46

CHAPTER 3	PROMINENCE ENVIRONMENT (O Engvold)	47

3.1   Introduction   47
3.2   Helmet Streamers   48
  3.2.1 Eclipse Photography   48
  3.2.2 Morphology   48
  3.2.3 Location of Current Sheet   51
  3.2.4 Brightness   52
3.3   Coronal Cavities   52
  3.3.1 Brightness and Structure   52
  3.3.2 Temperature and Density   53
3.4   Filament Channels   54
  3.4.1 Association with Neutral Lines   54
  3.4.2 Poleward Migration of Filament Channels   55
  3.4.3 Presence of Prominences   55
  3.4.4 Temperature and Electron Pressure   59
  3.4.5 Cool Matter in the Filament Channels   59
3.5   Prominence-Corona Transition Region   61
  3.5.1 Line Emission   61
  3.5.2 Empirical Modelling   61
  3.5.3 A Fragmented and Dynamic Transition Region   64
3.6   Prominences and Environment   66
  3.6.1 Magnetic Fields and Chromospheric Structure   66
  3.6.2 Association with Supergranulation Network   66
  3.6.3 Dynamics   68
  3.6.4 The Mass of Coronal Cavity and Prominence   68
  3.6.5 Coronal Voids - a Source of Prominence Mass?   71
3.7   Modelling of the Helmet Streamer/Prominence Complex   72
  3.7.1 Helmet Streamer and Cavity   72
  3.7.2 Magnetic Field Topology   73
  3.7.3 Siphon-Type Models   74
3.8   Conclusions   75

CHAPTER 4	OBSERVATION OF PROMINENCE MAGNETIC FIELDS (J L Leroy)	77

4.1   Historical Steps   77
4.2   Investigations Based on the Polarimetry of Spectral Lines   79
  4.2.1 Zeeman Effect   79
  4.2.2 Hanle Effect   83
  4.2.3 180° Ambiguity   85
  4.2.4 Instrumental Achievements   87
4.3   Indirect Magnetic Field Determinations   89
4.4   Magnetic Field at the Photospheric Level   90
4.5   Main Features of the Magnetic Field in Quiescent Prominences   93
  4.5.1 Field Strength   93
  4.5.2 Angle with Horizontal   99

CONTENTS

4.5.3	Angle with Prominence Axis	100
4.5.4	Magnetic Structure with Normal or Inverse Polarity	102
4.5.5	Homogeneity of the Field	107
4.6	Some Important Problems	109
4.6.1	Magnetic Field in Sub Arc Second Structures	109
4.6.2	Paradox of Fine Vertical Structures	109
4.6.3	Determination of Currents	110
4.6.4	Evolution of Prominence Magnetic Structure	111

CHAPTER 5   THE FORMATION OF SOLAR PROMINENCES (J M Malherbe)   115

5.1	Introduction	115
5.2	Overview of Observations	116
5.3	Main MHD Instabilities Involved in Prominence Formation	130
5.3.1	Radiative Thermal Instability	130
5.3.2	Resistive Instabilities	131
5.4	Steady Reconnection in Current Sheets	132
5.4.1	Incompressible and Compressible Theories	132
5.4.2	Unification of Different Regimes	133
5.5	Static Models	133
5.5.1	Condensation in a Loop	134
5.5.2	Condensation in an Arcade	134
5.5.3	Condensation in a Sheared Magnetic Field	134
5.5.4	Condensation in a Current Sheet	135
5.6	Dynamic Models: Injection from the Chromosphere into Closed Loops	135
5.6.1	Surge-Like Models	135
5.6.2	Evaporation Models	136
5.7	Dynamic Models: Condensation in Coronal Current Sheets	138
5.7.1	Numerical Simulations	138
5.7.2	Role of Shock Waves in Condensation Process	138
5.8	Unsolved Problems	140
5.9	Conclusion	141

CHAPTER 6   STRUCTURE AND EQUILIBRIUM OF PROMINENCES (U Anzer)   143

6.1	Introduction	143
6.2	Prominence Models	144
6.2.1	Global Structure	144
6.2.1.1	Two-Dimensional Equilibria	144
6.2.1.1.1	Models with Normal Magnetic Polarity	144
6.2.1.1.2	Models with Inverse Magnetic Polarity	149
6.2.1.1.3	Force-Free Fields	152
6.2.1.2	Quasi-Three-Dimensional Models	154
6.2.1.3	Support by Alfven Waves	154
6.2.2	Internal Structure and Thermal Equilibrium	155

	6.2.2.1 Hydrostatic Equilibrium	155
	6.2.2.2 Thermal Equilibrium	158
6.3	Concluding Remarks	164

CHAPTER 7 STABILITY AND ERUPTION OF PROMINENCES (A W Hood) 167

7.1	Introduction	167
7.2	Description of MHD Instabilities	168
7.3	Methods of Solution	170
	7.3.1 Normal Modes	170
	7.3.2 Energy Method	172
	7.3.3 Non Equilibrium	174
7.4	Effect of the Dense Photosphere	176
	7.4.1 Physical Arguments	176
	7.4.2 Ballooning Modes	177
7.5	Coronal Arcades	178
	7.5.1 Distributed Current Models - Eruptive Instability	178
	7.5.2 Localised Modes - Small Scale Structure	180
	7.5.3 Arcades Containing a Current Sheet	181
7.6	Thermal Stability	183
7.7	Resistive Instabilities - Tearing Modes	184
	7.7.1 Introduction	184
	7.7.2 Estimate of Tearing Mode Growth Rate	186
	7.7.3 Effect of Line Tying	187
7.8	Simple Model of Prominence Eruption and a Coronal Mass Ejection	188
7.9	Conclusions and Future Work	189

REFERENCES 193

INDEX 213

PREFACE

Prominences are amazing objects of great beauty whose formation, basic structure and eruption represent one of the basic unsolved problems in Solar Physics. It is now 14 years since the last book on prominences appeared (Tandberg-Hanssen, 1974), during which time much progress in our knowledge of the physics of prominences has been made, and so the time is ripe for a new text book which it is hoped will be a helpful summary of the subject for students, postdocs and solar researchers. Indeed, the last few years has seen an upsurge in interest in prominences due to high-resolution ground- and space-based observations and advances in theory. For example, an IAU colloquium was held in Oslo (Jensen et al, 1978), a Solar Maximum Mission Workshop took place at Goddard Space Flight Center (Poland, 1986), an IAU Colloquium is planned in Yugoslavia in September 1989 in prominences and it is expected that the SOHO satellite will be a further stimulus to prominence research.

In November 1987 a Workshop on the Dynamics and Structure of Solar Prominences was held in Palma Mallorca at the invitation of Jose Luis Ballester with the aim of bringing observers and theorists together and having plenty of time for in-depth discussions of the basic physics of prominences. It was appreciated that, as well as being important in their own right, the plasma-magnetic field interaction and radiative instability which are studied in prominences are of interest for similar processes in other astrophysical objects, such as cooling flows in galaxy clusters, structure in supernova remnants and cool clouds in other stars. Also, even though they have been studied scientifically for over a hundred years, many of their basic properties are still puzzling, such as how they form, what is their magnetic structure, what are the causes of their plasma flows, fine structure and feet, and why they erupt. In the relaxing atmosphere of a Mallorcan cafe the review speakers at this workshop decided to produce greatly extended (by a factor four) versions of their review talks to form the present book as an up-to-date account of the subject. After an introduction, three chapters deal with observations of the structure, environment and magnetic field of prominences, while three further chapters discuss the theory of prominence formation, structure and stability. The units that have been adopted are mks with magnetic fields measured in Gauss (a standard option in mks). It is hoped that this book will help the reader to share some of the fascination and excitement that the authors feel about prominence research.

E R Priest
St Andrews, June 1988

ACKNOWLEDGEMENTS

The authors gratefully acknowledge permission to reproduce the following copyright figures: Figs 1.1a, 2.1b, 2.2, 2.4a, 2.4b, 2.7, 2.8, 2.12, 2.15, 2.17, 3.11, 4.7, 5.1-5.8 (Meudon Observatory), Figs 1.1b ((High Altitude Observatory), Figs 1.2a, 1.5 (Big Bear Solar Observatory), Figs 1.2b, 2.14 (Sacramento Peak Observatory, AURA inc), Fig 1.3 (Swedish Observatory, La Palma), Fig 1.4 (R Tousey, Naval Research Laboratory), Figs 2.3a, 2.5, 3.14 (Pic du Midi Observatory), Figs 2.14, 2.16 (Pub. Astron. Soc. Japan), Fig 2.21 (L Gesztelyi, Debrecen Observatory), Fig 3.2 (T Tsubaki), Fig 3.9 (H Morishita), Fig 3.12 (E Hiei), Figs 3.17, 5.10 (T Forbes), Fig 3.20 (R M MacQueen), Figs 2.3b, 2.4c,d, 2.22 (Astron. Astrophys.), Figs 2.15, 2.17 (UVSP on Solar Maximum Mission), Fig 2.23 (X O Gu), Fig 3.15 (E Schmahl).

The editor is also most grateful to Jose Luis Ballester for organising the Workshop so well and for the secretarial help of Shiela Wilson that was undertaken with her usual efficiency, patience and good humour.

CHAPTER 1

INTRODUCTION TO QUIESCENT SOLAR PROMINENCES

E. R. PRIEST
Mathematical Sciences Department
The University
St Andrews KY16 9SS
Scotland

This first chapter is meant to set the scene for the more specialised chapters that follow. Its aim is to give a summary of prominence properties, to point out some of the major puzzles about them and to remind the reader of the MHD equations together with their physical significance.

Quiescent prominences are of great interest in their own right and are of wider importance both for Solar Physics and for many areas of Astrophysics. Their basic properties are not well understood at all, but they may be crucial for the mass and magnetic flux balance of the solar corona. Also they mark out large-scale magnetic polarity inversion lines in the photosphere, as well as the cores of large-scale magnetic arcades in the corona. The eruption of a prominence often produces a coronal mass ejection and sometimes a large two-ribbon flare. As far as astrophysics as a whole is concerned, prominences represent regions where magnetic fields are interacting with plasma in subtle ways, where dense plasma is being supported against gravity and where thermal instability is producing a cool condensation, and so by studying prominences in detail we can learn how these fundamental processes are likely to operate elsewhere in the universe.

1.1 Basic Description

1.1.1 DIFFERENT TYPES

Prominences are located in the corona but possess temperatures a hundred times lower and densities a hundred times greater than coronal values. In eclipse pictures, these cool, dense features appear bright at the limb (Figure 1.1a), but in $H\alpha$-photographs of the disc they show up as thin, dark, meandering ribbons called filaments (Figure 1.1b). They have been classified morphologically in several different ways, but there appear to be two basic types.

(1) A *quiescent prominence* is in its global appearance an exceedingly stable structure and may last for many months. It may begin life as a relatively small active-region (or plage) filament, which is located either along the magnetic inversion line between the two main polarity regions of an active region or at the edge of an active region where it meets a surrounding region of opposite polarity. Sometimes it may enter a sunspot from one side. As the active region disperses, the prominence grows thicker and longer to become a quiescent filament. It may continue growing for many months up to 10^6 km in length, and in the process it migrates slowly towards the nearest pole.

(2) *Active prominences* are located in active regions and are usually associated with solar flares. They are dynamic structures with violent motions and have life-times of only minutes or hours. There are various types, such as surges, sprays (which are probably erupting plage filaments) and loop prominences: both their magnetic field (about 100 G) and average temperature are higher than

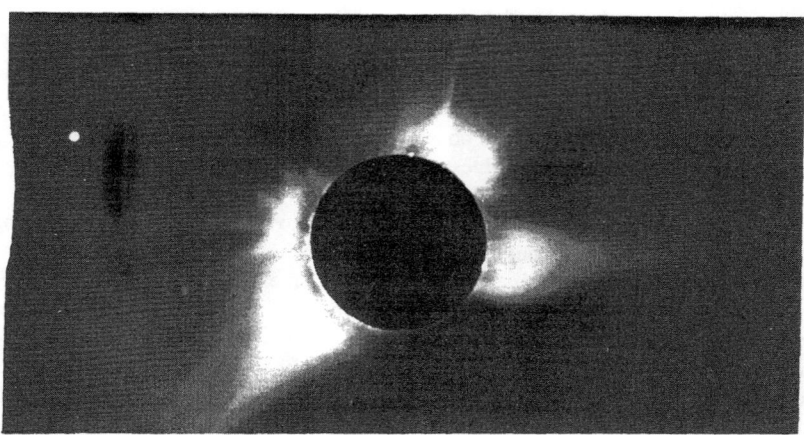

Figure 1.1(a) The disc of the Sun in $H\alpha$, showing many small plage filaments and two large quiescent prominences, one near the north-west limb and one at the north-east limb (courtesy Meudon Observatory). (b) White-light eclipse photograph of the corona, showing bright prominences with their overlying coronal cavities and helmet streamers (courtesy High Altitude Observatory).

for quiescent prominences. As far as this book is concerned we shall, however, only be dealing with their quiescent cousins (including plage filaments).

Quiescent prominences have captivated the solar observer for centuries. One was observed at an eclipse in the Middle Ages (1239) and described as a 'burning hole'; another in 1733 was called a 'red flame'. However, by the beginning of the nineteenth century, their existence had been forgotten. At the 1842 eclipse they were rediscovered, but the observers were so surprised that they did not give a reliable description: some even thought they were 'mountains' on the Sun. In 1860 they were photographed, and in 1868 spectroscopic techniques were introduced which led to the discovery of helium. Lyot's invention of the coronagraph in 1930 enabled him to observe prominences at the limb without waiting for an eclipse. Babcock found in 1955 that, when viewed on the disc, a filament always lies along a so-called polarity (or magnetic)-inversion line, where the line-of-sight magnetic field reverses its sign. Then, in 1957, Kippenhahn and Schluter put forward their classic model for the support of the dense prominence material against gravity by a magnetic field (Section 6.2.1). More recently, an alternative model was proposed by Kuperus and Raadu (1974), as described in Section 6.2.1.

1.1.2 PROPERTIES

A quiescent prominence is a huge, almost vertical sheet of dense, cool plasma surrounded by a hotter and rarer coronal environment. Its density ranges between 10^{16} m^{-3} and 10^{17} m^{-3} ($10^{10} - 10^{11}$ cm^{-3}), with the ratio of protons to neutral H atoms lying between about 1 and 10. The central temperature lies between 5000 K and 8000 K. The dimensions may have the following range: length 60,000 km to 600,000 km, height 10,000 to 100,000 km, thickness 4000 to 15,000 km. An active region prominence is typically a factor of three or four smaller than its mature quiescent form; its temperature is much the same, but its density is rather larger (10^{17} m^{-3}) and its height is at most 20,000 km. For further details see Chapter 2.

The magnetic field in quiescent prominences observed at the limb has a line-of-sight component (from the Zeeman effect) that varies from no observable field to 30-40 G. Tandberg-Hansen (1974) finds a mean value of 7.3G, with about half the observations in the range 3 to 8 G. Harvey (1969) gives a mean value of 6.6 G for 1967, while Rust (1967) gives 5 G for 1965 and finds that the magnetic field increases by roughly 50% over the height of the prominence. The average angle between the directions of the magnetic field and the long axis of a prominence is about 15°. Similar results have also been found by Leroy (1979) using the Hanle effect. Active-region prominences have higher field strengths, sometimes from 20 to 70 G, though possibly as much as 100 G; it appears that for this type the magnetic field may be aligned approximately with the filament, whereas for quiescent ones the field runs more across the filament (see Chapter 4).

The polarity of the magnetic field normal to the plane of the prominence has been a controversial topic (Section 4.5.4). We would like to suggest here the notation Normal (N) or Inverse (I) polarity. In the former case - for example in the Kippenhahn-Schluter model - the magnetic polarity is the same as a simple magnetic arcade in which the flux goes up on one side, passes through the prominence and goes back down to the photosphere on the other side. The latter case, exemplified by the Kuperus-Raadu model, has the opposite polarity in the prominence for the same signs of the photospheric flux as before. We prefer this notation to K-S (Kippenhahn-Schluter) and K-R (Kuperus-Raadu) when simply referring to the sign of the prominence flux rather than specific models. We also prefer it to the notation P (potential-like) and NP (non-potential-like), since all quiescent prominence magnetic fields are greatly sheared and therefore highly non-potential.

1.1.3 DEVELOPMENT

Prominences change slowly in overall shape with a lifetime between 1 and 300 days. Low-latitude quiescent prominences possess a mean lifetime of about 2 rotations (50 days), whereas those at high latitudes have an average duration of 5.1 rotations (140 days). The formation of a filament within an active region takes typically a few hours or a day (Chapter 5). A quiescent filament may appear more slowly (over the course of a day or so), either between two nearby active regions, or at the boundary of an active region, or in a remnant active region (Martin, 1973); but always the birth takes place at a polarity inversion line. Another condition for the formation of many filaments within an active region or between adjacent active regions is that the $H\alpha$ fibrils first align themselves end-to-end along a path called a filament channel, which eventually becomes a filament. Once the filament is formed the fibrils on either side are found to be directed roughly parallel to the filament. This alignment of fibrils suggests that the magnetic field is directed approximately along the filament. Eventually, a quiescent prominence disappears by either slowly dispersing and breaking up, or erupting, or flowing down to the chromosphere.

As a prominence migrates towards a pole, it is stretched more and more by the action of differential rotation into an east-west direction, while its width and height remain relatively constant. Prominences tend to occupy zones about 10° polewards of the sunspot latitudes, which move towards the equator as the solar cycle progresses. They are also located in polar zones, where they are oriented nearly parallel to the equator and sometimes form a polar crown around a polar cap at about latitude 70°. The polar zones form about 3yr after spot maximum and migrate towards the poles; at about the next maximum they reach the poles and are accompanied by the reversal in polarity of the polar magnetic fields.

1.1.4 STRUCTURE

Often a prominence reaches downwards towards the chromosphere in a series of regularly spaced feet, which resemble great tree trunks. These feet are often located at supergranule boundaries and are joined by huge arches (Figure 1.2). Within a prominence there is much fine structure in the form of vertical threads of length 5000 km and diameter about 300 km or less (Figures 1.2, 1.3); material appears to stream continually and slowly down these threads and down the arches into the chromosphere at speeds of only 1 km s^{-1}, which is much less than the free-fall speed although they may not represent true mass motions. The resulting loss of mass is immense and would drain the prominence in a day or so if it were not being replenished somehow. Hardly any motion along the axis of a quiet quiescent filament is observed, unless it interacts with a sunspot, but active-region filaments often show matter flowing along the axis into a sunspot.

In eclipse and coronagraph pictures, one finds a region of reduced density, known as a coronal cavity, surrounding a prominence (Figure 1.1); it can also be seen in soft X-ray photographs as a region of reduced intensity, but the mass deficit may be insufficient by an order of magnitude to account for the prominence mass. A hot, closed coronal arcade or arch is present above and around the cavity, and above that there lies a helmet streamer; it may live for several months and has a broad base (up to a solar radius in diameter) with a pointed top at 1 or 2 solar radii above the limb. At transition region temperatures the large-scale region around a prominence is found to possess upflows at 6 to 10 km s^{-1}. The prominence-corona interface between the cool threads and the hot corona is extremely thin (at most only a few hundred kilometres) and so possesses an exceptionally high temperature gradient; the electron pressure there is about 2×10^{-3} N m^{-2} (2×10^{-2} dyne cm^{-2}), which is about a quarter of the quiet-Sun transition region pressure but about the same as

Figure 1.2. Prominences in $H\alpha$ at the limb (courtesy H Zirin, Big Bear Solar Observatory and Sacramento Peak Observatory, AURA inc).

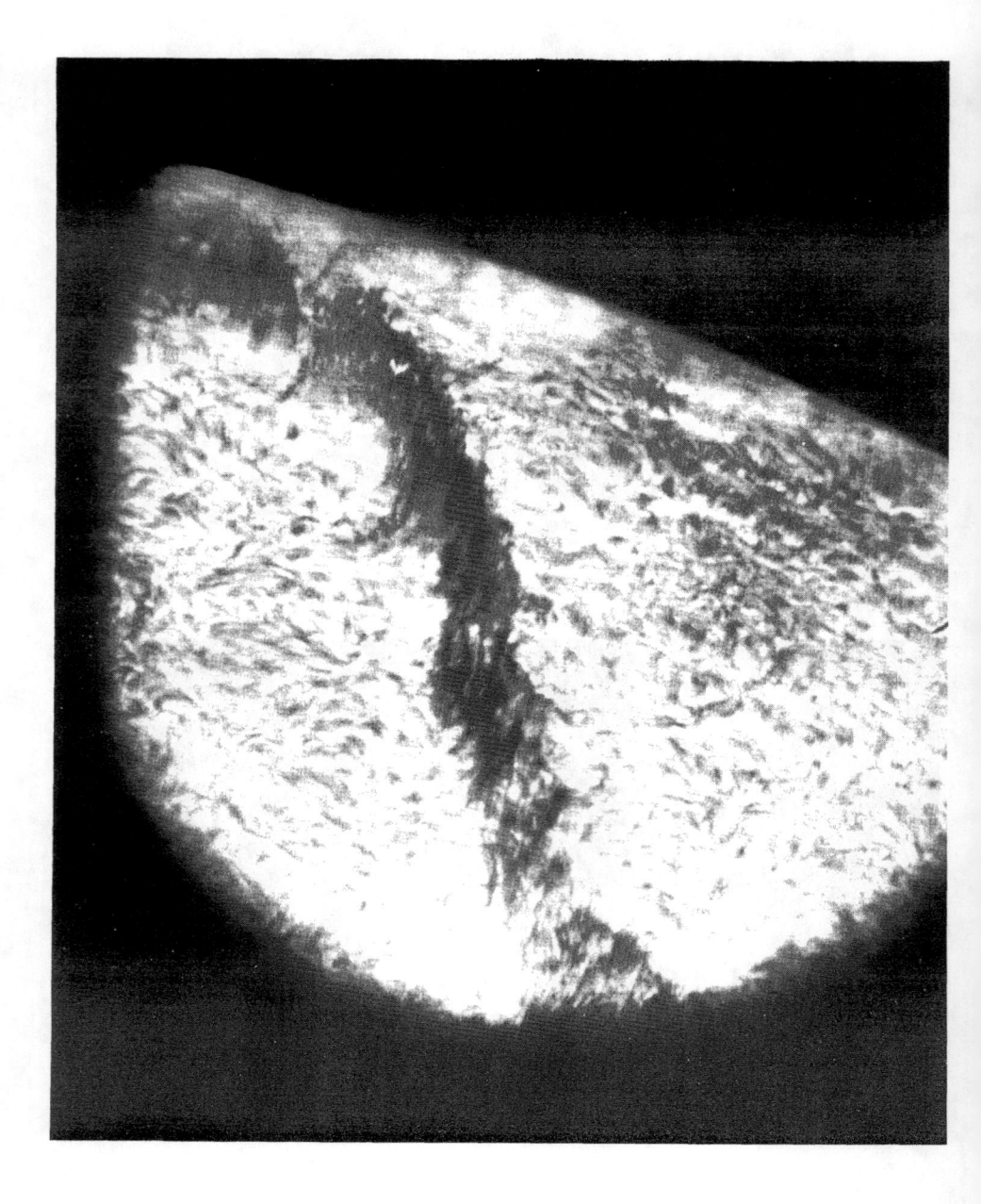

Figure 1.3 High-resolution $H\alpha$ picture of a prominence on the disc showing fibril structure (courtesy O Engvold, Swedish Observatory, La Palma).

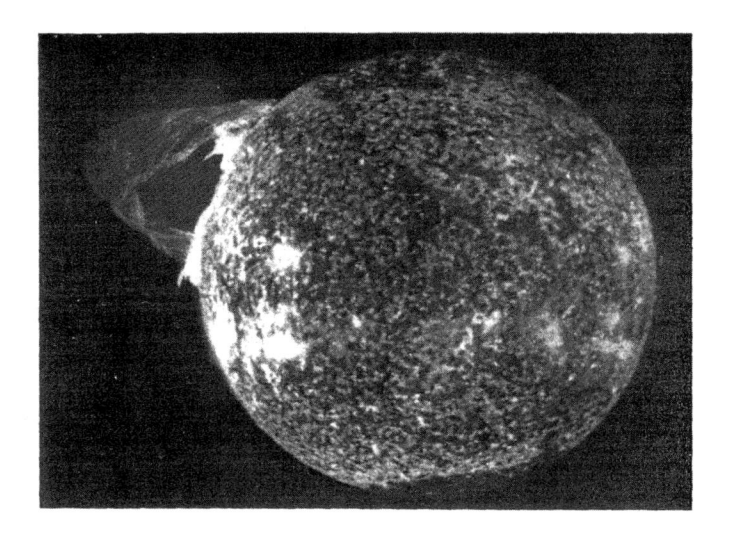

CHAPTER 1: INTRODUCTION TO QUIESCENT SOLAR PROMINENCES 7

Figure 1.4 Eruption of a quiescent prominence in He II 304 (courtesy R Tousey, Naval Research Laboratory)

as the prominence pressure (see Chapter 3).

1.1.5 ERUPTION

Active-region and quiescent prominences can become *activated* and exhibit several types of large-scale motion. For example, a prominence may become larger and darker (when viewed on the disc as a filament) or brighter (when viewed at the limb). At the same time, there may be an increase in turbulent (or helical) motion or flow along the filament. This type of activation sometimes fades away after an hour or so and sometimes it leads to an *eruption*, as described below. In other cases, prominence material may drain away from the summit along a curved arc at speeds of 100 km s^{-1}. Another type of activation is called a *winking filament*, when the prominence performs a damped oscillation for 2 to 5 oscillations with a period of 6 to 40 min; it is initiated by the passage of a shock wave from a distant large flare.

At some stage in its life, an active-region filament or quiescent filament may become completely unstable and erupt, especially once it exceeds about 50,000 km in height. It ascends as an *erupting prominence* (Figures 1.4, 1.5) and eventually disappears; some of the material escapes from the Sun altogether while some descends to the chromosphere along helical arches.

In two-thirds of cases, the prominence reforms in the same place and with much the same shape over the course of 1 to 7 days. The eruption of an old *quiescent prominence* is referred to as a *disparition brusque* (sudden disappearance); it starts with a slow rising motion at a few km s^{-1} and may take several hours. It is accompanied by an X-ray brightening and occasionally by the appearance of $H\alpha$ flare ribbons. Usually the cause of the eruption is a mystery, but sometimes it may be initiated by a disturbance from an emerging flux region or a flare. The eruption of an *active-region filament* is much more rapid and takes about half an hour or less; while it is still ascending at high speed, a two-ribbon $H\alpha$ flare begins and, in this case, the filament generally reforms after

OCTOBER 10, 1971

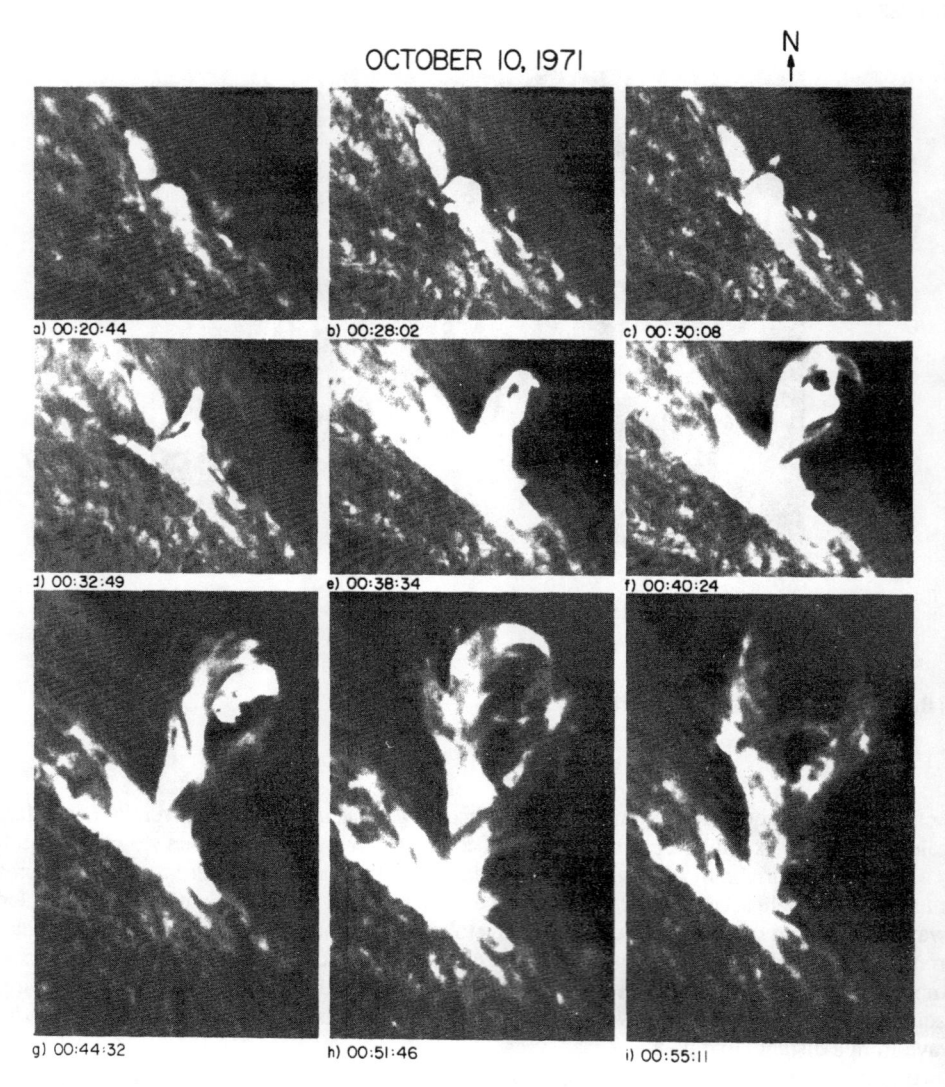

N ↑

a) 00:20:44   b) 00:28:02   c) 00:30:08
d) 00:32:49   e) 00:38:34   f) 00:40:24
g) 00:44:32   h) 00:51:46   i) 00:55:11

Figure 1.5 Eruption of a plage filament in $H\alpha$ and the accompanying flare brightening (courtesy Big Bear Solar Observatory).

only a few hours. For further details see Sections 2.4 and 7.5.

1.2 Basic Equations of Magnetohydrodynamics

The fundamental equations for the plasma velocity (**v**), magnetic field (**B**), plasma density (ρ),

CHAPTER 1: INTRODUCTION TO QUIESCENT SOLAR PROMINENCES

electric current (j), plasma pressure (p) and temperature (T) are the induction equation

$$\frac{\partial \mathbf{B}}{\partial t} = \nabla \times (\mathbf{v} \times \mathbf{B}) + \eta \nabla^2 \mathbf{B}, \tag{1.1}$$

the equation of motion

$$\rho \frac{d\mathbf{v}}{dt} = -\nabla p + \mathbf{j} \times \mathbf{B} + \rho \mathbf{g}, \tag{1.2}$$

the mass continuity equation

$$\frac{d\rho}{dt} = -\rho (\nabla \cdot \mathbf{v}), \tag{1.3}$$

the energy equation (for optically thin radiation)

$$\frac{\rho^\gamma}{\gamma - 1} \frac{d}{dt} \left(\frac{p}{\rho^\gamma} \right) = -\nabla \cdot (\kappa \nabla T) - \rho^2 Q(T) + \frac{j^2}{\sigma} + h\rho, \tag{1.4}$$

where $\nabla \cdot \mathbf{B} = 0$,

$$\mathbf{j} = \nabla \times \mathbf{B}/\mu \tag{1.5}$$

is Ampere's law, μ is the magnetic permeability and

$$p = \frac{k_B}{m} \rho T = n k_B T = \frac{\mathcal{R}}{\tilde{\mu}} \rho T \tag{1.6}$$

is the perfect gas law in terms of the mean particle mass (m), the Boltzmann constant (k_B) and the particle number density (n). The significance of equation (1.1), in which $\eta = (\mu \sigma)^{-1}$ is the *magnetic diffusivity* and σ is the electrical conductivity, is that changes in magnetic field strength are caused by transport of the magnetic field with the plasma (as represented by the first term on the right), together with diffusion of the magnetic field through the plasma (the second term on the right). In order of magnitude, the ratio of the first to the second term on the right is the *magnetic Reynolds number*

$$R_m = \frac{v_0 l}{\eta} \tag{1.7}$$

in terms of a typical plasma speed (v_0) and length-scale (l) for magnetic variations. For length-scales comparable with the size of typical coronal structures, the magnetic Reynolds number is enormous (say, $10^6 - 10^8$) and so diffusion is negligible and the magnetic field is effectively frozen to the plasma. It is only in intense current concentrations such as narrow threads or sheets, where l is extremely small (kilometres or even less), that diffusion and therefore reconnection can take place. A related dimensionless parameter is the *Lundquist number*

$$S = \frac{v_A l}{\eta} \tag{1.8}$$

where

$$v_A = \frac{B}{(\mu \rho)^{1/2}} \tag{1.9}$$

is the *Alfvén speed*. It may be written as the ratio

$$S = \frac{\tau_d}{\tau_A} \tag{1.10}$$

of the *magnetic diffusion time* ($\tau_d = l_0^2/\eta$) to the *Alfvén travel time* ($\tau_A = l_0/v_A$) and so measures the relative importance of diffusion and advection of magnetic flux.

Equation (1.2) shows that the plasma is acted on by forces due to a plasma pressure gradient, the magnetic field and gravity. The magnetic force may be rewritten using (1.5) as

$$\mathbf{j} \times \mathbf{B} = -\nabla \left(B^2/(2\mu) \right) + (\mathbf{B} \cdot \nabla)\mathbf{B}/\mu, \qquad (1.11)$$

which represents the sum of a magnetic pressure force acting from regions of high to low magnetic pressure ($B^2/(2\mu)$) and a magnetic tension force acting towards the centre of curvature of curved magnetic field lines. A useful parameter is the *plasma beta*

$$\beta = \frac{p}{B^2/(2\mu)} \qquad (1.12)$$

which measures the ratio of the pressure gradient to the magnetic force. It may also be written as

$$\beta = \frac{2}{\gamma} \frac{c_s^2}{v_A^2}, \qquad (1.13)$$

where γ is the ratio of specific heats and

$$c_s = \left(\frac{\gamma p}{\rho} \right)^{1/2} \qquad (1.14)$$

is the *sound speed*.

In equation (1.4), which is appropriate for the optically thin corona, entropy changes are caused by the terms on the right, namely thermal conduction, optically thin radiation, ohmic heating and a small-scale heating term, which is assumed to be proportional to density with a constant of proportionality h. κ is the (tensor) coefficient of thermal conduction. The optically thin radiation is proportional to density squared and has a temperature dependence (Q(T)) which possesses a maximum at about $10^5 K$ and a minimum at about $10^7 K$. At lower temperatures the radiation may become optically thick, conduction becomes radiative rather than Coulomb and the small-scale heating term disappears. Also, in some applications viscous terms are added to the momentum and energy equations.

A few features of the solutions to the MHD equations are mentioned below. More details can be found in the books by Cowling (1976), Roberts (1967) and Priest (1982).

1.2.1 MAGNETOHYDROSTATICS

Slowly changing structures, such as prominences, are in approximate equilibrium under a balance between various forces. In particular, if flow speeds are much smaller than both the sound speed, the Alfven speed and the free-fall speed $(gl)^{1/2}$, equation (1.2) reduces to

$$0 = -\nabla p + \mathbf{j} \times \mathbf{B} + \rho \mathbf{g}. \qquad (1.15)$$

Since the $\mathbf{j} \times \mathbf{B}$ force is perpendicular to $\mathbf{B}$, the component of (1.15) along the magnetic field may be written

$$0 = -\frac{dp}{dz} - \rho g \qquad (1.16)$$

CHAPTER 1: INTRODUCTION TO QUIESCENT SOLAR PROMINENCES

when gravity acts in the negative z-direction. From the perfect gas law (1.6)

$$\rho = \frac{mp}{k_B T}, \qquad (1.17)$$

and so ρ may be eliminated between (1.16) and (1.17) and the resulting equation integrated to give

$$p = p_0 exp - \int_0^z \frac{dz}{\Lambda}, \qquad (1.18)$$

where

$$\Lambda = \frac{k_B T}{mg} \qquad (1.19)$$

is the *pressure scale-height*. If, in particular, the temperature is uniform, (1.18) becomes

$$p = p_0 exp(-z/\Lambda), \qquad (1.20)$$

and so the pressure decreases exponentially with height along each magnetic field line. The scale-height is typically 100 km in the photosphere, 500 km in the chromosphere and 100,000 km in the corona, so that as one rises up above the solar surface the pressure falls off extremely rapidly at first and later much more slowly.

In the particular case when $\beta \ll 1$ and the vertical distance under consideration is much less than Λ/β, (1.15) reduces to

$$0 = \mathbf{j} \times \mathbf{B} \qquad (1.21)$$

and the magnetic field is said to be *force-free*, where

$$\mathbf{j} = \nabla \times \mathbf{B}/\mu \quad \text{and} \quad \nabla \cdot \mathbf{B} = 0.$$

Very little is known about general solutions to the nonlinear equation (1.21) in spite of its apparent simplicity (Low, 1982). Even less is known about (1.15), where the pressure and density terms provide a coupling to the energy equation. Only rather simple solutions have been investigated for prominences so far.

1.2.2 WAVES

For a uniform plasma in a uniform magnetic field ($\mathbf{B}_0$), equations (1.1)-(1.4) may be linearised by writing variables in the form

$$\mathbf{B} = \mathbf{B}_0 + \mathbf{B}_1 exp\, i(\omega t - \mathbf{k} \cdot \mathbf{r}) \qquad (1.22)$$

and retaining only terms linear in $\mathbf{B}_1/\mathbf{B}_0$. Here ω is the wave frequency and $\mathbf{k}$ is the wavenumber vector in a direction θ, say, to the magnetic field $\mathbf{B}_0$. When the frequency is so high that terms on the right of (1.4) are negligible, the variations are adiabatic and the modes become particularly simple.

For example, in the absence of gravitational and pressure effects ($g = p_0 = 0$), there are two distinct magnetic modes, namely the *shear Alfven wave*, with a dispersion relation

$$\omega = k v_A \cos\theta, \qquad (1.23)$$

and the *compressional Alfven wave*, for which

$$\omega = k v_A. \qquad (1.24)$$

When the plasma is not cold ($p_0 \neq 0$) and the magnetic field is negligible we have a sound wave

$$\omega = kc_s \tag{1.25}$$

In the presence of a magnetic field this mode couples to the compressional Alfven wave to give slow and fast *magnetoacoustic waves*.

Gravity gives rise to a gravity wave with dispersion relation

$$\omega = N\sin\theta_g, \tag{1.26}$$

in terms of the Brunt frequency

$$N = (\gamma - 1)^{1/2} \frac{g}{c_s} \tag{1.27}$$

and the inclination (θ_g) of the direction of propagation (k) to the force of gravity. In general, this too couples to the magnetoacoustic modes.

The above simple picture is made much more complex by considering a non-uniform basic state (see Roberts, 1985). A hierarchy of different modes appears when the medium is structured in the form of, for example, an interface, a slab or a flux tube. A further complication arises when the disturbances are no longer linear. For example, the magnetoacoustic modes can steepen to form shock waves, with a *slow shock* causing the magnetic field to rotate towards the shock normal as it passes and a *fast shock* making it rotate the other way and so increasing the field strength rather than decreasing it. Also, intermediate waves (or finite-amplitude Alfven waves) can cause the tangential component of the magnetic field to reverse its sign while the magnitude of the field is unchanged.

1.2.3 INSTABILITIES

Instabilities may be discovered by seeking perturbation solutions to (1.1)-(1.4) about an equilibrium $\mathbf{B}_0$ (in general non-uniform) in the form $\mathbf{B}_0 + \mathbf{B}_1 exp\,\omega t$.

The magnetic field is found to modify some of the basic plasma (or fluid) instabilities (see Chapter 7) such as:

Rayleigh-Taylor instability, which has a growth-rate $\omega = (gk)^{1/2}$ and occurs when a dense plasma is supported against gravity on top of a rarer plasma (Section 7.2);

Kelvin-Helmholtz instability, which takes place when one plasma streams over another one;

radiative instability, which is driven by the optically thin radiative loss term $\rho^2 Q(T)$ in the energy equation (Section 7.6).

In addition there are certain new instabilities that exist when a magnetic field is present such as:

kink instability of a magnetic flux tube which is sufficiently twisted (Section 7.2);

resistive instability of a sheared magnetic field in which magnetic field lines break and reconnect (Section 7.7).

1.3 Prominence Puzzles

Our knowledge of quiescent prominences has increased significantly over the past ten years but there are many basic features which are still either uncertain or just not understood at all. The remaining chapters aim to give an in-depth account of our current understanding of prominences, the first three dealing with observations of the interior, environment and magnetic fields, while the last three describe theories of formation, equilibrium and stability. It is hoped that the reader will be encouraged to help answer some of many outstanding questions so that we may increase our understanding of these fascinating creatures.

Chapter 2 describes and classifies the different types of prominence and stresses the importance of filaments as tracers of the general magnetic field. Spectroscopic diagnostics in photospheric, chromospheric and EUV lines are used to deduce the density, temperature, microturbulence and velocity in and around prominences. The activation due to instabilities together with the consequences of an eruption are described. Outstanding questions include: what are the relation and relative roles of the global and small-scale structure; what causes the prominence feet and how do the supergranule cells and feet evolve; what is the relation between the prominence and the underlying concentration of magnetic flux in the photosphere; what is the global mass balance and circulation pattern in a prominence and what drives it; what are the detailed properties of and what gives rise to the fine-scale fibril structure?

Chapter 3 summarises observations of the prominence environment including filament channels, coronal cavities and the overlying coronal arcades and helmet streamers. It also describes the narrow interface between the prominence and the corona. A significant fraction of the length of a long-lived photospheric polarity inversion line possesses overlying prominences, but what determines which portions are with and without a prominence? Is magnetic shear the crucial parameter? How much more cool material is there than is observed in $H\alpha$? What is the energy balance in the transition region, and how important is enthalpy flux there? Is the transition region completely dynamic? What is the role of flows in prominence formation and maintenance? What is the relationship of flows observed in different lines and can the doppler shifts be confidently interpreted as bulk flows? What is the nature and cause of the various oscillations seen in prominences?

Chapter 4 describes the measurements of magnetic fields based on the Zeeman effect and the Hanle effect. Most quiescent prominences have fields between 3 and 30 Gauss, which are close to horizontal and make a small angle ($\approx 20°$) with the prominence long axis. For low-lying prominences (height less than 30,000 km) the magnetic polarity is Normal (N) with a field component across the prominence sheet that is consistent with that of a simple arcade based on the underlying photospheric field, but for high prominences the polarity is Inverse (I) with the opposite orientation. Although the magnetic field looks rather homogeneous at the present resolution, a big question is how to measure subarcsecond fields and how such fields are related to chromospheric structure. Also, what is the three-dimensional magnetic structure of a prominence and in particular what is the nature of feet? Why are there different magnetic types of prominence and how are they related? How does the magnetic configuration evolve?

Chapter 5 gives an account of the static and dynamic models that have been proposed for prominence formation. Static models include thermal condensation in a loop, an arcade, a sheared magnetic field or a current sheet. Dynamic models include injection from below (as a surge or an evaporation) and condensation in a current sheet aided by fast-mode shocks which may be produced by magnetic reconnection. At the present stage, many of these theories are plausible, and so they need to be refined and compared in detail with observations in order to determine which are responsible for different types of prominence. Surprisingly, the conditions for onset of thermal condensation under realistic boundary conditions are still not well understood. Furthermore, a consensus does not exist as to whether condensation from the corona or injection from the chromosphere is most likely to form a prominence.

Chapter 6 discusses theories for the structure and equilibrium of a prominence. Two-dimensional models for global magnetic equilibria include the classical Kippenhahn-Schluter and Kuperus-Raadu models which now need to be superceded by more realistic three-dimensional models incorporating an external highly sheared configuration. For the internal structure it is necessary in future to couple the magnetohydrostatics and nonLTE radiative transfer and also to include dynamic effects in the energy balance. Furthermore, there are as yet no well-developed models of fine-scale prominence threads.

Finally, Chapter 7 describes the basic theory for magnetic and thermal stability and applies it to coronal and prominence structures including loops, arcades and Kippenhahn-Schluter models. The variational and normal modes methods are compared and the importance of choosing the most realistic boundary conditions is stressed. Outstanding questions include: what is the global instability or nonequilibrium mechanism that is responsible for prominence eruptions; what causes fine structure in prominences - is it a localised ballooning mode, for example; what causes a thermal disappearance without eruption?

CHAPTER 2

OVERALL PROPERTIES AND STEADY FLOWS

B.SCHMIEDER
Département Solaire et Planétaire
Observatoire de Paris-Meudon
92195 Meudon Principal Cedex
France

Since the publication of the book *Solar Prominences* by Tandberg-Hanssen (1974), prominences have been the subject of many theoretical and observational studies, as discussed and summarized in several reviews (Hirayama 1985, Poland 1986, Malherbe 1987). This chapter presents recent progress, mostly from an observational point of view in the understanding of the formation, structure, support and instabilities of solar prominences. Results of spectroscopic diagnostics (velocity, temperature and density) will be stressed.

2.1 Basic Properties

2.1.1 DESCRIPTION AND CLASSIFICATION

Prominences (or filaments when they are observed on the disk) define cool structures ($6,000^0$K- $10,000^0$K) imbedded in the solar corona (Figure 2.1). In $H\alpha$, they appear as bright features on the limb and dark features on the disk. The observations suggest that they are supported against gravity by a horizontal magnetic field. In fact the term prominence is used to describe a variety of objects ranging from relatively stable structures with lifetimes of many months, to transient phenomena that last a few hours or less. Prominences have been classified in several ways. As was pointed in Chapter 1, this book concerns only quiescent prominences. Sudden ejection of cool material (surges, sprays and ejecta), which may appear in regions without any preexisting filaments, will not be considered because their mechanisms of formation are different from prominences. Quiescent prominences can be classified in two principal classes as was mentioned by the first observer of prominences, Secchi (1875). He divided them into 2 main classes : quiescent and plage filaments.
Quiescent prominences are located outside active regions (Figure 2.2). They migrate towards latitudes higher than 45^0 where they are called polar prominences. Their characteristic dimensions are 200Mm long, 50Mm high and 10Mm wide. Their lifetimes can reach several months. They are anchored in the photosphere at footpoints which are periodically separated (30Mm) and may be due to a plasma collapse (Nakagawa and Malville 1969).

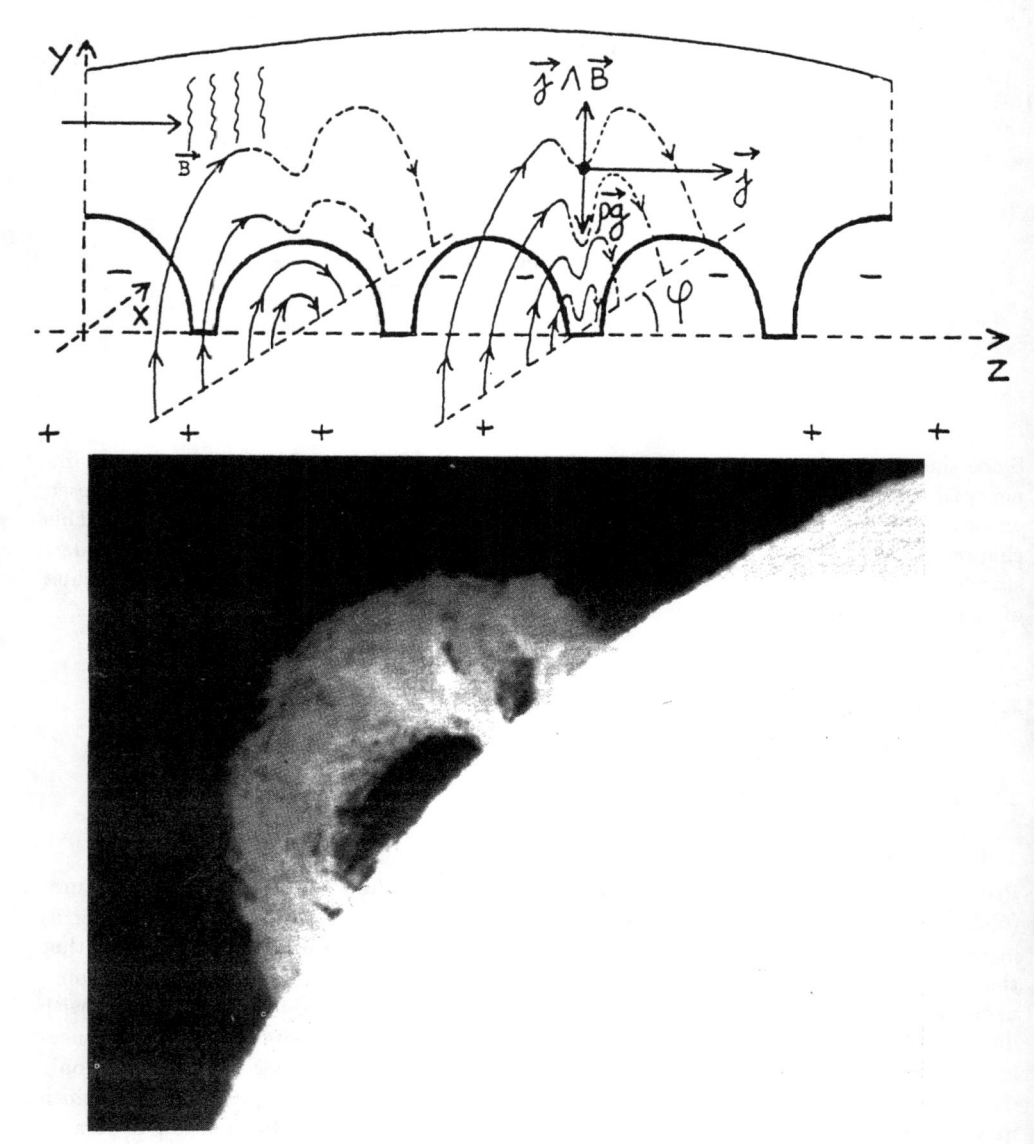

Figure 2.1: (a) Schematic of a prominence imbedded in the corona overlying an inversion line of the photospheric magnetic field(+/-). Solid lines represent the arch boundaries of the filament, while light and dashed lines represent magnetic field lines supporting the prominence. (b) prominence with an arch structure (Meudon Spectroheliogram)

They look at the limb like arches or vertical structures in $H\alpha$ according to perspective effects (Figure 2.2), and are situated inside a coronal cavity visible in coronal lines Fe XIII (Noëns 1987, see Chapter 3).

Observations on the disk in $H\alpha$ show a quiescent filament structure with footpoints connected by large arches visible only when the thickness is sufficient. In CIV, filaments are also darker than the surrounding region (Figure 2.3) and the footpoints are clearly visible. The arch is generally too faint and instead, a classical transition region is seen (Simon et al 1986).

Plage filaments or active region filaments are located in active regions. They are lower and not so extended (10Mm high, 50Mm long). They lie along a neutral line of the photospheric magnetic field and appear to consist of different extended low-lying loops limited by footpoints, anchored either near a sunspot or between faculae (Figure 2.4a,c).

Near their footpoints filaments are sometimes invisible in $H\alpha$ and in CIV when the faculae brightness is too large or the spatial resolution too low. 30% of active region filaments become quiescent. Their time evolution and support will be treated in sections 2.3.2 and 4.6.4.The physical parameters of these filaments will be analysed in section 2.2, the instabilities thermal or dynamical in section 2.3, and some other dark $H\alpha$ structures such as post-flare loops will be mentioned.

2.1.2 FINE STRUCTURE IN $H\alpha$

Quiescent filaments have two different structural time scales, a long one (3 rotations on average) related to the general loop shape and a short one (4-10 min) related to the fine structure whose visibility is limited by the spatial resolution (Figure 2.5a).

Simon et al (1986a) have recently described how the well organized quiescent prominence that we have referred to in section 2.1.1 can appear as clusters of small scale loops oriented in all directions when the seeing is good (Pic du Midi). This is clearly visible when Figures 2.3a and 2.5a, representing the same filament with two different seeings, are compared. Dunn (1960) stated that the small features in a prominence are narrow, certainly less than 300 km. The observations made by Engvold (1976) at the limb also show knots of averaged half-width comparable with the instrument spatial resolution ($\sim$ 400 km, see Figure 2.6).

Démoulin et al (1987) detect fine threads 7000 km long and with a half-width of 700 km on MSDP observations obtained at Pic du Midi. Such a fine structure is continuously evolving with a typical time scale of about 8 min (Engvold 1976).

The thermal stability of fine structure has been investigated by several authors. The generally accepted global models (a cool core with a hot sheath envelope) are valid only as a first approximation. If we look at high-resolution images of a prominence observed in different lines formed at various temperatures, the structure is composed of fine threads heterogenous in temperature and density . Several different models have been suggested for them :

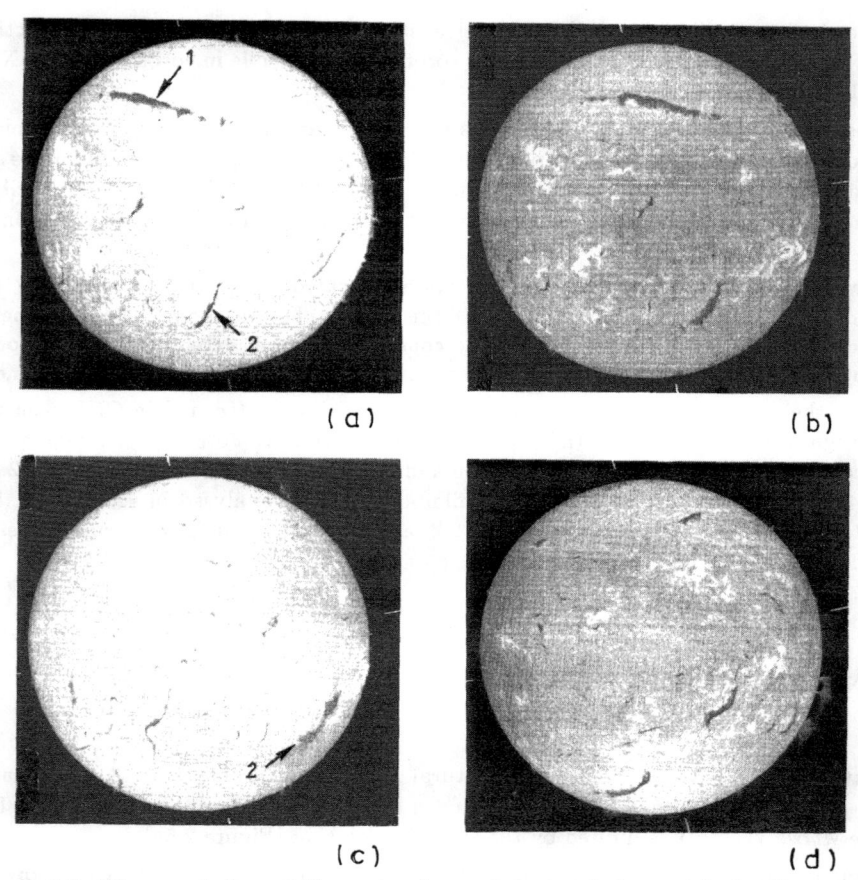

Figure 2.2: Time evolution of filaments observed during 8 days with the Hα spectroheliograph on (a) July 26, (b) 28, (c) 30 and (d) August 2 1980. The polar crown prominence labelled 1 in (a) is no longer visible 2 days after (b). The filament labelled 2 looks like a prominence at the limb (d).

- hot sheaths around cool cores (Kundu 1986, Gary 1986)
- isothermal threads (Poland and Tandberg-Hanssen 1983)
- threads with longitudinal temperature gradients along the magnetic field (with radiative losses balanced by conductive flux).

The third model was proposed by Schmahl and Orrall (1986) looking for a good fit of the differential emission measure derived from the EUV line intensity observed with Skylab. Morover, Rabin (1986) shows how classical models determined by thermal conduction along magnetic fields fail because the very thin transition region does not radiate enough to provide the increase of differential emission measure towards lower temperatures. He

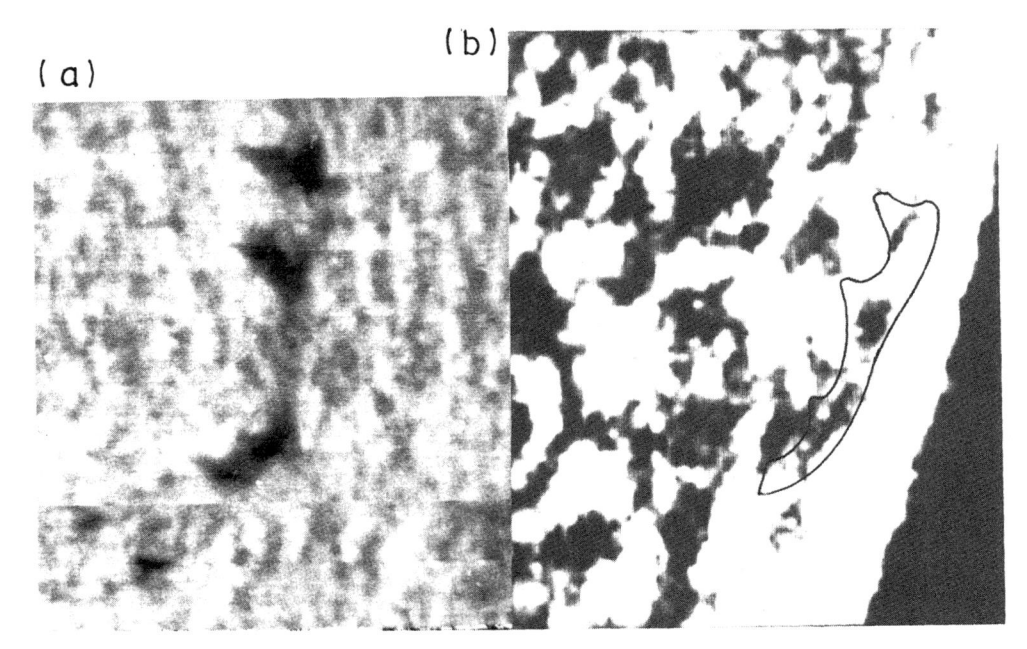

CHAPTER 2: OVERALL PROPERTIES AND STEADY FLOWS   19

Figure 2.3: Hα and CIV intensity maps of a quiescent filament (Oct.17 1984) observed on the disk with (a) the MSDP at Pic du Midi, and (b) the UVSP aboard the SMM spacecraft. Between the footpoints, the classical CIV transition zone around the arch is visible.

therefore proposes a model with 1000 threads along the line of sight. Démoulin et al (1987) follow this up by deducing from a theoretical study of parallel and orthogonal conduction , characteristic lengths and time scales using a chess-board model. They find that parallel conduction leads to instability for a length larger than 45 km but this mechanism is too efficient with too small a time scale. The orthogonal conduction time scale reaches an hour and can be comparable with the typical lifetime of prominence threads.
It is now clear that the thermal stability must be taken into account simultaneously in MHD models, which is necessary for future modeling of prominences.

2.1.3 EVOLUTION OF FILAMENTS DURING THE SOLAR CYCLE

A study of long time scales shows the importance of filaments as tracers of the general solar magnetic field. Hα filaments are located between large regions of opposite polarity. During their lifetime, they become aligned in an East-West direction due to differential rotation. Following their trajectory through the solar cycle, large-scale azimuthal magnetic structures have been determined (Ribes 1986).
    This study has been complemented by sunspot rotation measurements. The migra-

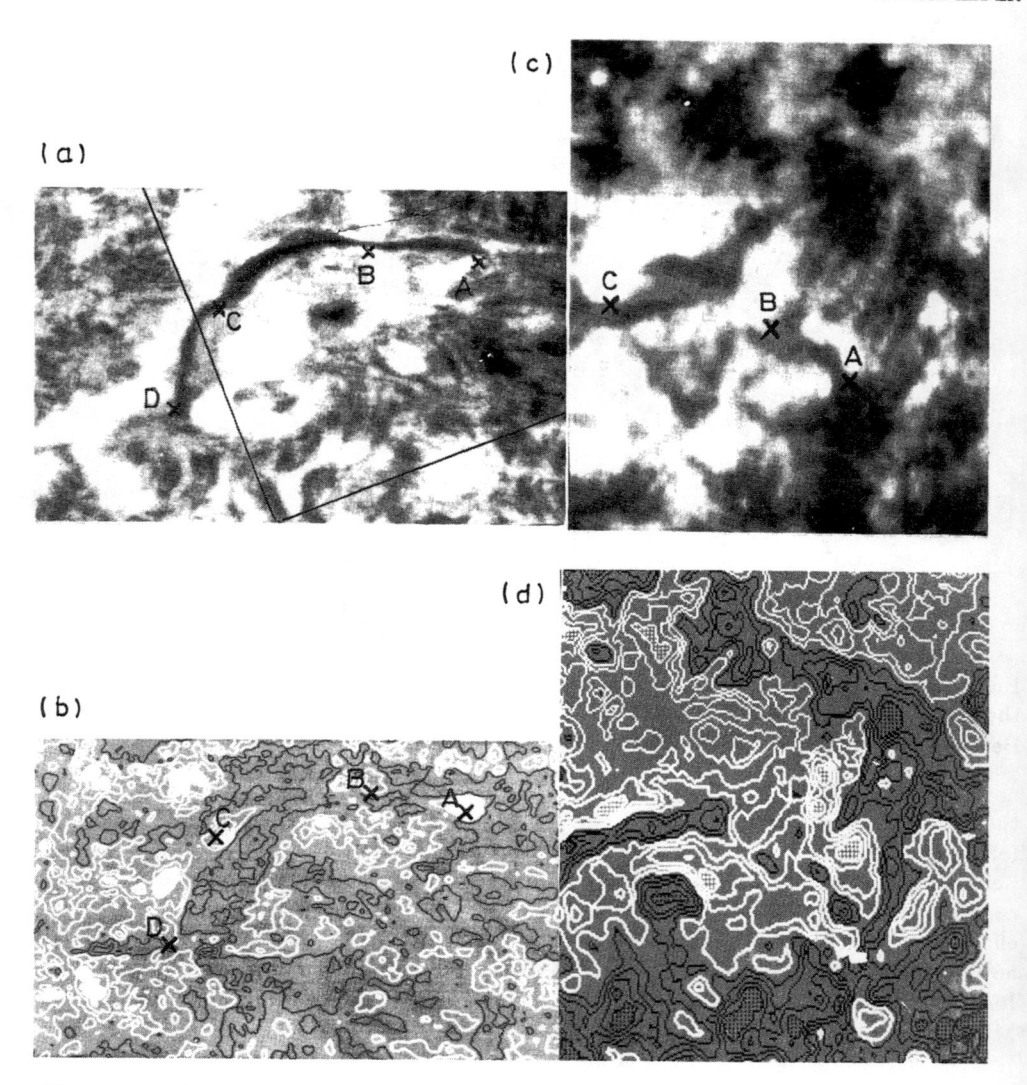

Figure 2.4: Plage filament observed on Sept.30, 1980, (a) H_α (MSDP operating at Meudon) and (c) CIV (UVSP aboard SMM) intensity, (b) and (d) H_α and CIV Dopplershifts. The box in (a) represents the field of view of (c). Blueshifts (black contours) are observed between the footpoints (A,B,C,D) and redshifts above (white contours).

tion of sunspots from 40^0 to the equator has been well known since the XIX th century. However, recently, using the collection of spectroheliograms obtained at Meudon since 1917, (after digitalization and computation with sophisticated code), Ribes et al (1985) found a complex zonal meridional circulation following the trajectory of young sunspots. Only

(a) (b)

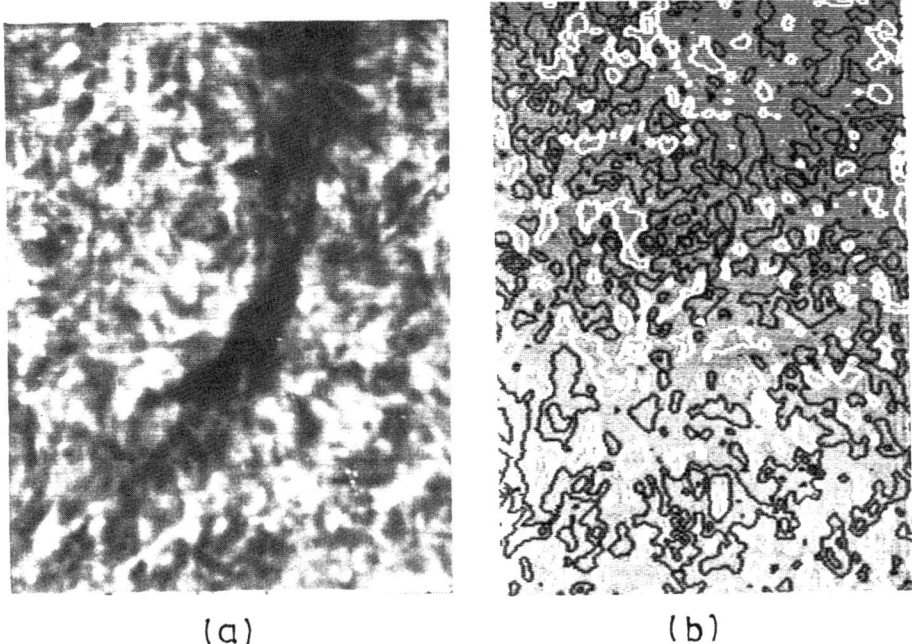

Figure 2.5: Fine structure in H_α observed in a filament at Pic du Midi with the MSDP spectrograph, on Oct.15 1984 (a) I and (b) V at H_α ±0.25Å. The velocity pattern is relatively confused in the filament channel.

young sunspots were considered because they are still anchored deep in the photosphere and follow convective motions, so that their rotation period is shorter than that of old ones. The borders of unipolar regions defined by the filaments lie at the latitudes where the meridional circulation reverses. This coincidence suggests the existence of East-West oriented magnetic rolls. which appear first at $40°$ of latitude at the beginning of the cycle and migrate slowly to the poles with a mean velocity around 20 ms^{-1} (Figure 2.7).

They could be of convective origin as they seem to be associated with a pulsation of the solar envelope(Ribes, Laclare 1988). They can be theoretically interpreted as the convective response of the toroidal field in the framework of the dynamo theory (Galloway and Jones 1988) . It is not clear , however , that the rolls represent a unique system of convective motions in the generation of magnetic activity as suggested by Snodgrass and Wilson (1987) . Glatzmaier(1985), Gilman and Miller (1986) also proposed theoretical models with meridional cells. In any case, it is obvious that filaments and the solar convective zone are connected. A study of large-scale structure could lead to a better understanding of prominence structure and formation.

Some surface features of the sun such as the young spots do not follow the well determined differential rotation law . Some long-lived filaments also present irregularities in rotation (D'Azambuja 1948). Owing to the behaviour of these filaments (7 to 10 rotations) singularities in the solar rotation have been detected (Soru-Escaut et al 1984,1985).

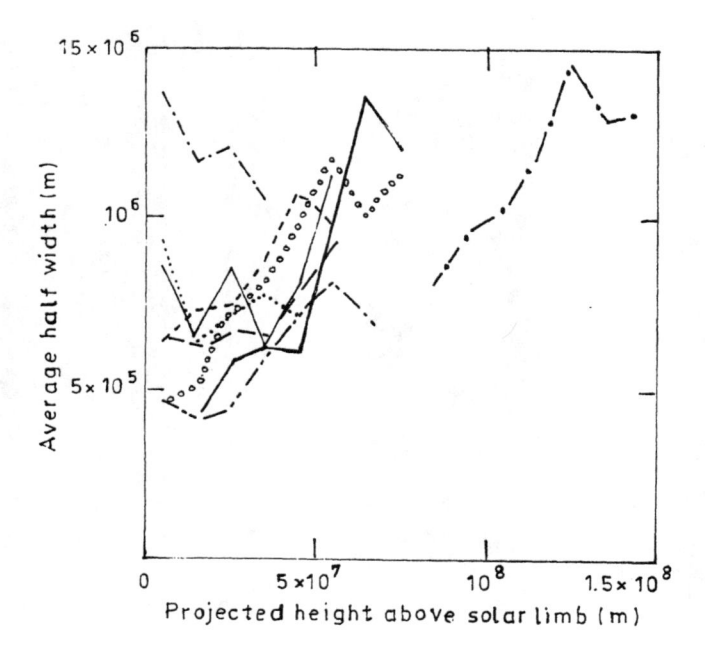

Figure 2.6: Averaged half widths of knots or threads, observed in prominences above the limb, versus altitude (Engvold 1976).

Limited solar areas of rigid rotation called *pivot points* have been defined(Figure 2.8) and recently a relationship has been found between rigidly rotating filaments, new emerging active centres and flare onsets (Martres et al 1987, Mouradian et al 1987). The behaviour of filaments as a function of the solar cycle was also investigated by Gnevyshev and Makarov (1985), Makarov (1984).

2.2 Physical Characteristics: Density and Temperature

2.2.1 DENSITY AND IONIZATION DEGREE

The generally accepted values of electron density (about 10^{16} to 10^{17} m^{-3}) and degree of ionization ($n_p/n_1 \leq 1$ to 3) are now questionable (see the review of Vial 1986). Depending on the methods used, high or low values may be found (Figure 2.9 top). Bommier et al (1986), for example, have measured density values 10 times smaller ($\sim 10^{16}$ m^{-3}). Using fourteen prominences of high altitude(> 30 000 km) observed with the Pic du Midi corono-

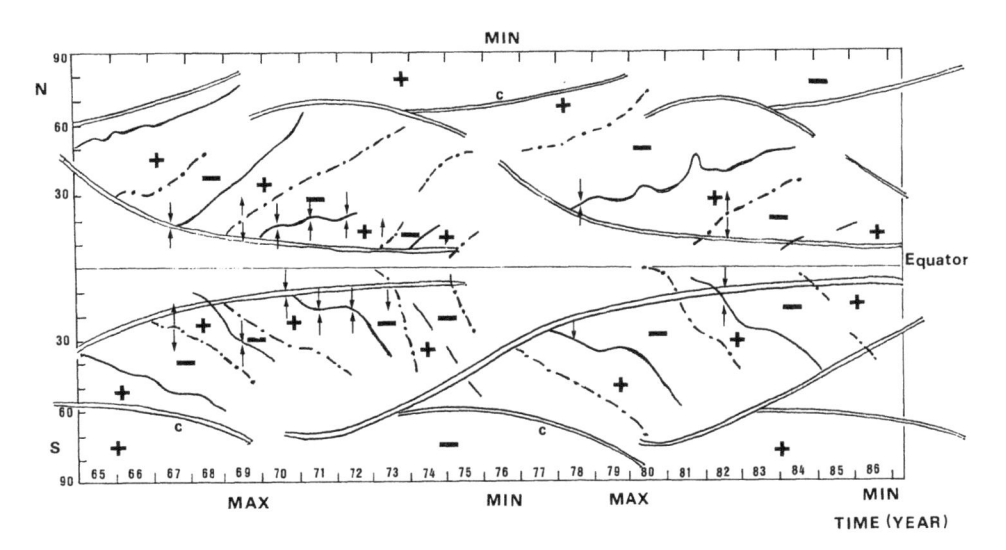

CHAPTER 2: OVERALL PROPERTIES AND STEADY FLOWS 23

Figure 2.7: Azimuthal rolls observed over 20 years. The time is plotted horizontally and the latitude vertically. The arrows determinated by the migration of young spots indicate the meridional circulation. Large scale magnetic structures (dashed and continuous lines) are drawn using the trajectories of H_α filaments (Ribes, Laclare 1988).

graph, the electron density has been derived from quasi-simultaneous linear polarisation measurements in the He I D_3 and hydrogen H_β lines. The linear polarization of D_3 and H_β is due to the resonant scattering of the underlying photospheric radiation field and is modified by the local magnetic field, which leads to a depolarisation and a rotation of the polarisation (the Hanle effect).

The polarization of H_β is in addition modified by the depolarizing effect of collisions with the electrons, protons and He^+ ions of the medium (Sahal-Bréchot 1984, Bommier 1986). Such a method leads to the determination of the electron density (1.10^{15} m^{-3} to 4.10^{16} m^{-3}) and the magnetic field vector(see chapter 4). Low density values ($5.7\ 10^{15}$ m^{-3}) were also determined by Koutchmy et al (1983) and by Stellmacher et al (1986) using observations made during an eclipse using H_β and the Balmer continuum. Landman (1985a), however, deduced electron densities larger than 10^{17} m^{-3} from the ratio of metallic lines (Na I, Sr II and Ba II) and high density values were confirmed by Hirayama (1985) using the Stark effect of high Balmer lines. He determined average electron densities of $10^{17.02}$ m^{-3} for five *hedgerow* quiescent prominences and $10^{16.48}$m^{-3} for two *curtain* like prominences. Hirayama suggests that $n_e \leq 10^{16.0}$ m^{-3} may be expected for the fainter part of a quiescent filament. Bright prominences obviously have larger n_e values. These values are based on an evaluation of the filling factor which is connected directly to the fine structure of the filament. Besides, Hirayama(1986) pointed out that the values of n_e do not depend only on the type of prominence observed but also on the method used. The degree of ionization of hydrogen is also uncertain, and varies according to the authors from 0.05 to 1 (Landman

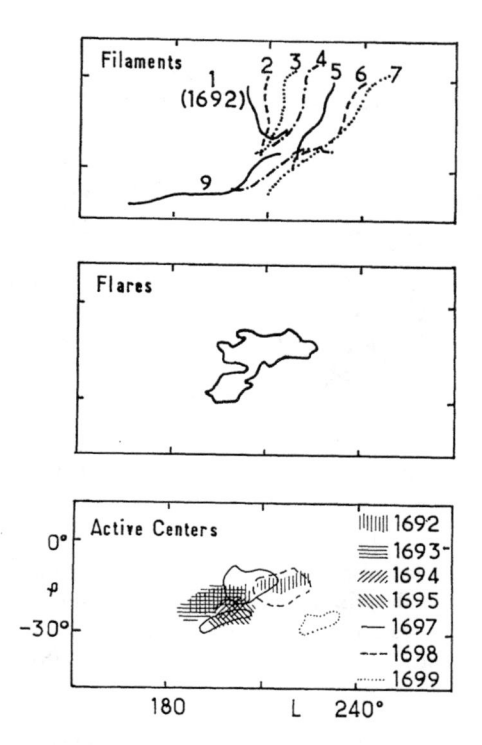

Figure 2.8: (a) *Pivot Points* indicated by trajectories of filaments during 9 solar rotations taken from Meudon Synoptic maps. (b) Flare distribution area located at the "Pivot point" location. (c) Positions of new active centres

1984, Hirayama 1985) with a larger value than 1 at the edges of prominences. Vial (1982a) obtained degrees of ionization of the order of 3 using the ratio between L_α and Ca K lines. These disagreements are important for the physics of prominence structure. With typical values ($n_e = 10^{17}$ m^{-3}, $n_{HII}/n_{HI} = 1$, T=7000°K, B=10G) β is equal to 10^{-2} while with Bommier's (1986) values $\beta = 10^{-3}$ and Landman's values $\beta = 1$. If the magnetic field is only 3G, the extreme values become respectively equal to 10^{-2} and 10. In the first case the magnetic forces are the most important forces and the prominences can be condensed in situ in the corona; in the second case gas pressure forces dominate, the density of the prominence is equal to that of the corona. The prominence could not exist because of the large pressure scale height incompatible with the altitude of the prominence.

2.2.2 NLTE MODELS

During the last 15 years several NLTE models, theoretical or semi-empirical have been pro-

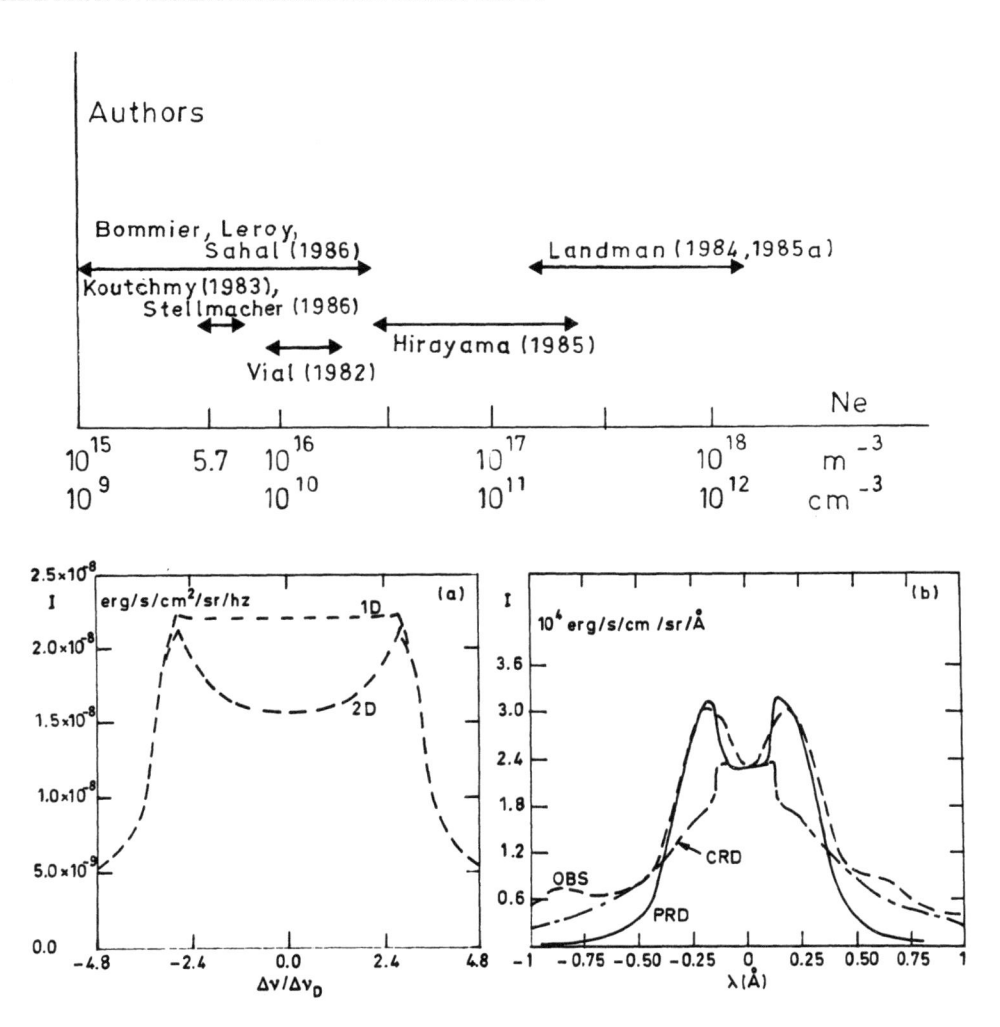

CHAPTER 2: OVERALL PROPERTIES AND STEADY FLOWS 25

Figure 2.9: (at the top) Electron density values in prominences according to various authors. (at the bottom) Reversed L_α profile obtained : (a) with 2D geometry; (b) with partial frequency redistribution (PRD) in NLTE models compared with observational profile (OBS) and profile obtained with complete redistribution (CRD).

posed for prominences (it is difficult to discriminate between the two types of model as is commonly done for stellar atmospheres). The equations of radiative transfer and statistical equilibrium must here be solved simultaneously (Mihalas 1978). Hydrogen emission lines may serve as indicators of temperature and density structure together with the Lyman continuum and can provide information about the excitation and ionization conditions in the prominence plasma. Before 1970, most work was devoted to the study of hydrogen Balmer lines but now with the advent of UV space observations (Skylab 73,OSO 8, SMM 80) it has become possible to fit hydrogen lines of the Lyman series such as L_α and L_β. More complete calculations using the technique of complete linearization have been performed by Poland et al (1971), Heasley et al (1974), Heasley and Mihalas (1976), Heasley and Milkey (1976,1978). These authors start with classical one-dimensional slab models, generally isothermal, of quiescent prominences irradiated by photospheric, chromospheric and coronal radiation. Special attention was given to the geometry used: Vial (1982b) used a 2D geometry and obtained reversed L_α profiles (Figure 2.9a bottom) while Fontenla and Rovira (1985), Morozhenko (1978) and Zharkova (1984) proposed thread models.The threads were defined by their pressure and central temperature by the first authors but were supposed isothermal by the others, who were principally interested in thread interactions. Some empirical models avoiding the isothermal assumption were recently developed by Zhang and Fang (1987). They reproduce observed profiles and deduce the variations of source function, electron density, and microturbulence. Another attempt was made by Heinzel et al (1987,1988), incorporating partial frequency redistribution instead of a complete redistribution and taking into account a thin transition region with high pressure around the prominence (Figure 2.9b bottom). This approach allows a more precise calculation of the Doppler dimming in L_α (lower intensity) and the introduction of a transition region in the model provides L_β profiles comparable with the observations. Heinzel and Rompolt (1987) also evaluated recently the influence of radial velocity on hydrogen lines in moving prominences and found how the Doppler dimming increases with the velocity.

2.2.3 TURBULENT VELOCITY AND ELECTRON TEMPERATURE

The non-thermal velocity (NTV) called microturbulence and electron temperature T_e are known to be 6 kms^{-1} and 7000^{0}K respectively if we consider average values (Hirayama 1971,1985). A relationship between NTV and T_e has been pointed out by Vial et al (1980) (Figure 2.10).

Nevertheless some problems are not resolved. The results obtained by Hirayama have shown that both T_e and NTV increase outward monotonically, which is supported by non-LTE model calculations (Heasley and Mihalas 1976, Zhang and Fang 1987). From the observational point of view, however, Zhang et al (1987) found a temperature 5000^{0}K at the edges of the prominence and 9000^{0}K in the central region and a decreasing microturbulence from the centre to the edge of prominences (Figure 2.11).

Engvold and Brynildsen (1986) also observed such a behaviour of T_e and NTV using IR lines. The short lifetime of structures could be due to rapid changes in temperature or density. Zhang et al explain the discrepancy by suggesting that the models have a distribution of the physical parameters along the line of sight assuming an axis of symmetry in the prominence. However, that is not generally the case in observed prominences because of

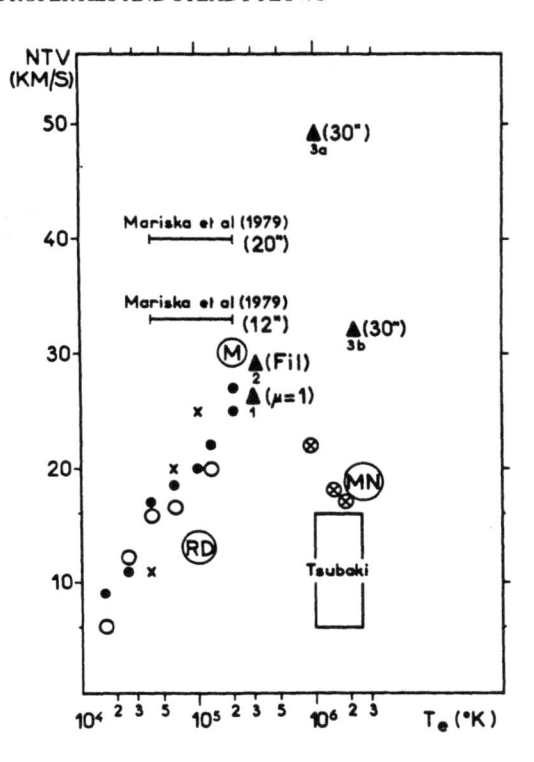

Figure 2.10: Relationships between non-thermal velocity (NTV) and electron temperature T_e using different observations (see Vial 1980 for details)

unresolved fine structures. Moreover, Mouradian and Leroy (1977) did not find any radial variation of temperatures in prominences. New computations of microturbulence are being made using the broadening of the observed H_α line by an inversion line process (Cram 1986, Mein and Schmieder 1988). Microturbulence is found to be associated with relatively large gradients of vertical velocity in $H\alpha$ and CIV (Schmieder et al 1987) and corresponds to a region of activity. An increase of microturbulence in a region of anomalous profiles could indicate a scattering of velocities along the line of sight. In regions of instability , for example, NTV could reach 30 kms^{-1} (Figure 2.12).

The observed temperature of 7000^{0}K is well reproduced coming from incident Lyman continuum radiation if a thin slab model is considered (Hirayama 1985). But for a thicker model which is closer to the observations a small temperature is expected. To explain a temperature higher than 7000^{0}K, an extra heating source is needed. Further work is certainly necessary here.

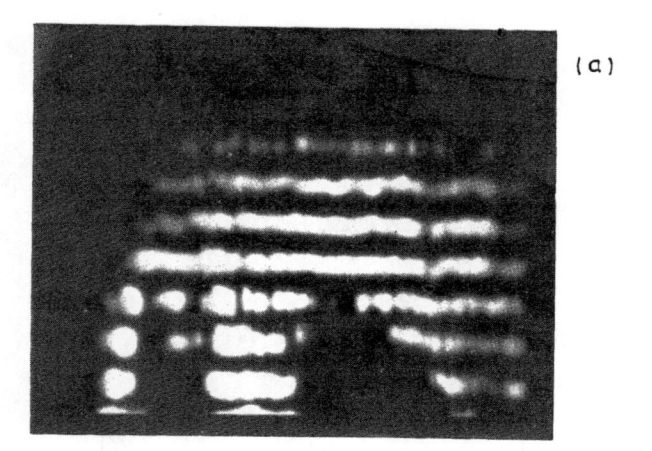

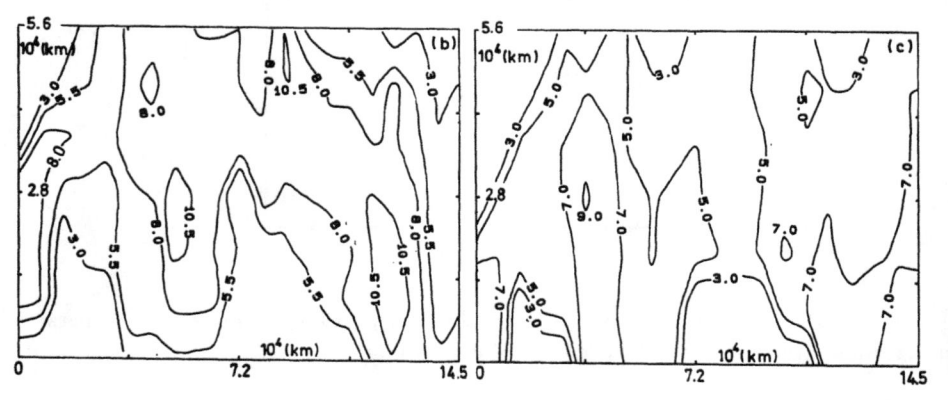

Figure 2.11: (a) Prominence observed with the 13.5 m spectrograph of McMath telescope at Kitt Peak using H_α and Ca II K lines (Zhang et al 1987) (b) electron density (c) microturbulence.

2.3 Velocity Field and Mass Flux

It is clear that the basic state of filaments is not static but dynamic. We describe first the different types of instruments used to observe the velocity field and discuss the accuracy of the measurements. Therefore, we review velocity field observations below, in and around filaments.

2.3.1 INSTRUMENTATION

The analysis of velocity in prominences is based on different types of observation:

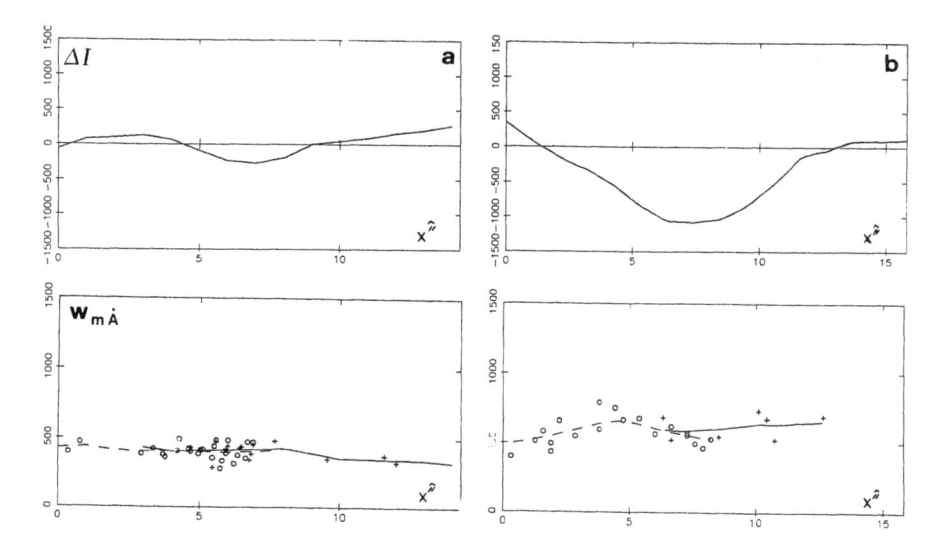

Figure 2.12: Cuts of intensity microturbulence through filaments observed with the MSDP. The microturbulence is larger through the unstable filament (b) than in the stable filament (a). The thermal width is around 300-400 mÅ. 600 mÅ corresponds to 25 km^{-1}, assuming a temperature around 8000 K.

(1) *Filtergraph*

Evolution of prominences at the limb or on the disk are classically observed using a filter technique (coronagraphs, Big Bear Solar Tower, Meudon 3λ heliograph, Astronomical Observatory of Wroclaw University). Temporal sets of filtergrams in Hα or movies provide apparent motions of chromospheric material. Special attention should be paid to the interpretation of such motions in terms of plasma velocities. Apparent motions in prominences observed at the limb may be due to variations in the degree of ionization of material at different levels, or to wave propagation through the atmosphere .

Observations in Hα wings give information on the existence of large velocities, for example $\Delta\lambda = 0.75$ Å corresponds to a velocity of 35 kms^{-1} at $\lambda = 6543$Å, if the profiles are assumed to be gaussian, symmetric and not broadened. These methods lead to qualitative results.

(2) *Spectrograph*

The spectrograph technique is best for obtaining quantitative Doppler-shift values. Generally the wavelength scale is calibrated and the velocities are accurate. The following instruments observe prominences in several wavelengths in chromospheric lines such as Hα,Ca II, H and K lines, He lines : the Oslo Solar telescope (Engvold et al 1978), Sac Peak Solar Vacuum Telescope (Engvold and Keil 1986) with a 100x100 CCD camera, Hida Solar

Tower in Kyoto(Kubota and Uesugi 1986), SSHG at Yunnan Observatory (Gu et al 1984) , MacMath telescope at Kitt Peak observatory (Zhang et al 1987), or in UV lines such as Lα, CaII, Mg II and CIV : the spaceborne OSO8 (Vial et al 1979)and HRTS spectrographs (Dere et al 1986) . Using a scanning process , a region can be observed where the field of view is limited by the length of the slit ($\sim$ 1') and the number of scans covering up to $\sim$ 1'. The HRTS spectrograph has an exceptionally long slit (one solar radius) but few scans are made . The best spatial resolution along the slit is about 0.5" using ground based telescopes (but often reduced to 1" or 2" because of the seeing quality), and 1" with spaceborne instruments. In the scanning direction the spatial resolution depends on the step used (3" to 7") . The spectral resolution reaches 10^{-2}Å with ground-based instruments and only 5×10^{-2}Å with spaceborne instruments.

(3) *Multi-Channel Subtractive Double Pass Spectrograph*

MSDP spectrographs provide 2D observations in several wavelengths (n)covering a line profile. This technique was first developed at the Meudon Solar Tower using the $H\alpha$ line with n=9 and an elementary field of view equal to 1'x8' (Mein 1977). Other MSDP's are being, or will be, developed at Pic du Midi (n=11 , 45"x5'), on the VTT and on the THEMIS telescopes operating on the Canary islands (n=22 , 2 lines simultaneously , CCD camera).

The 2D character of the MSDP has the advantage that it can cover quickly an entire active region. There is no relative distortion due to seeing between the n wavelengths (recorded simulateously). The profile of the line is reconstructed using a code taking into account photometry corrections and calibrations.

(4) *Lambdameter technique*

This technique uses a spectrograph where two detectors are placed symmetrically on the profile of the line , recording two intensities I_1 and I_2. $I_1 + I_2$ gives the total intensity, $I_1 - I_2$ the Dopplershift. This technique allows the use of CCD or reticon cameras, eliminating the use of film. It is operated using differential profiles of Zeeman components of photospheric lines on the Meudon (Rayrole 1981) and the Sac Peak magnetographs. Photospheric Doppler-shifts, independent of line profiles, are obtained (Martres et al 1981) which is not the case with Babcock magnetographs such as Kitt Peak magnetograph (Malville and Schindler 1981).

The UVSP spectrograph aboard the SMM satellite was working in several modes, e.g. large dopplergrams in CIV lines (4'x4') were obtained using 2 detectors, like simultaneous filtergraphs,and small scale images (1'x1') with a quasi-lambdameter technique. They were produced with the grating at a slightly different location for successive images , which permits one to obtain crude line profiles determined by 4 points at each pixel using the Double Doppler technique called DDD (Simon et al 1982). Velocities are more accurate but determining the zero velocity by averaging over the whole field of view (4'x4') is doubtful. The error is estimated to range between 4 and 10 kms^{-1} according to: Mariska (1986), and Athay et al (1983), if the general redshift of the transition region is taken into account.

2.3.2 $H\alpha$ PROFILE ANALYSIS

The contrast profile $C(\Delta\lambda)$ is a typical quantity defining the prominence from the observed intensity $I(\Delta\lambda)$ and the background chromospheric intensity $I_0(\Delta\lambda)$. It can be approximated by the following expression as first proposed by Beckers (1964):

$$C(\Delta\lambda) = \frac{I(\Delta\lambda) - I_0(\Delta\lambda)}{I_0(\Delta\lambda)}$$

$$= [\frac{S}{I_0(\Delta\lambda)} - 1][1 - e^{-\tau(\Delta\lambda)}]$$

$$\tau(\Delta\lambda) = t_0 exp[-(\frac{\Delta\lambda - V_c\lambda_{H\alpha}/c}{\Delta\lambda_c})^2]$$

with c being the speed of light. The source function S is assumed constant in the filament, V_c is the velocity in the filament, $\Delta\lambda_c$ is the Doppler width of the $H\alpha$ line, τ the optical thickness of the filament. Of necessity, caution must be used in interpreting such contrast profiles, considering the simplifying assumptions made, namely that we can adequately describe I_0 and that S is independent of wavelength and constant throughout the filament. The four parameters S, t_0, v_c, $\Delta\lambda_c$ are interconnected as pointed out in several studies (Grossmann-Doerth and Von Uexkull 1971, Mein et al 1985) The interpretation of the $H\alpha$ profiles needs some assumptions, principally the definition of a reference profile. The classical standard MSDP process uses as reference profile, a mean profile obtained over the observed field of view, excluding active regions and using a least square method . Doppler shifts are obtained at a given wavelength ($\Delta\lambda$) in the $H\alpha$ profile by the displacement of the middle of the chord , 2x$\Delta\lambda$ long, measured in the mean profile and compared to a profile measured in the filament. The measurements are relative. If the profile of the $H\alpha$ line in the filament is symmetrical, the Doppler shift is independent of $\Delta\lambda$. Kubota and Uesugi (1986) define an absolute wavelength for the center of $H\alpha$ using neighbouring points in the filament and a visual method which is very sensitive to the assumed gaussian shape of the $H\alpha$ profile.These two methods suppose also that the the optical thickness of the filament is large.

If the profile is asymmetrical, the profile may be interpreted as the combination of a stationary reference profile emanating from the lower chromosphere and a moving absoption profile from an overlying cloud. Different methods have been used, called *Cloud* model methods(Grossmann-Doerth and Von Uexkull 1971, Schmieder et al 1988, Heinzel 1987) or *Differential Cloud* Model methods of first and second order (Mein and Mein 1988, Mein and Schmieder 1988) . The differential method of first order allows the chromospheric background I_0 to vary over the field of view; using the second order variations along the line of sight of S, radial velocity and microturbulence can be derived. These methods are used to detect velocity shear in prominences.

2.3.3 VERTICAL MOTIONS

The analysis of photospheric dopplergrams indicates that velocity field amplitudes ($< 0.350 kms^{-1}$) *under filaments* in photospheric layers are reduced compared with those

Table 2.1: – Velocity field in prominences and filaments

Techniques	Upward motion (km/s)	Downward motion (km/s)	Horizontal motion (km/s)	Authors
Prominence on the limb				
H_α image		downward 15-35		Dunn 1960 / Engvold 1976
Mg II-Ca II profile			20	Vial et al 1979
O VI profile			8	Vial et al 1980
CaII, D_3 profile			30	Engvold et al 1978
Filament on the disk				
H_α profile	(6)[1]	(5.3)		Kubota 1980
		0.92		Kubota and Uesugi 1986
	0.5	footpoint		Martres et al 1981
			5	Malherbe et al 1983
	(5)	(5-10)		Schmieder et al 1985b
				Simon et al 1986
$H_e I$	< 3			Engvold and Keil 1986
C IV	5	footpoint		Schmieder et al 1984a
Dopplergram			< 5	Simon et al 1986
				Schmieder et al 1988
	(15)	(10)		Klimchuk 1986
			±10	Athay et al 1985
				Engvold et al 1985
C IV profile	5-6	footpoint		Dere et al 1986
Si IV profile			6-10	Lites et al 1976

(1) The numbers between brackets are maximum values.

CHAPTER 2: OVERALL PROPERTIES AND STEADY FLOWS   33

outside. This is related to diminished oscillations. The velocity vector is not horizontal in the photosphere and the velocity lines are roughly perpendicular to the direction of the filament (Figure 2.13). Also the size of velocity cells (< 15") is smaller than outside the filament region (Martres et al 1976, 198( ).

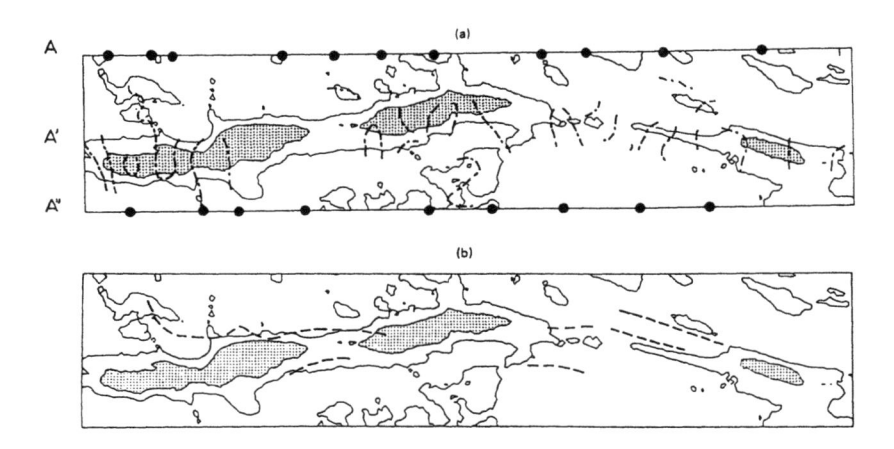

Figure 2.13: Inversion lines of radial velocity (a) in the photosphere (Fe I 6502Å) and (b) in the chromosphere ($H\alpha$). The $H\alpha$ intensity contours of the filament are drawn on the (a) map (Martres et al 1981).

There is certainly a relationship between the photosphere and filaments since they are observed to be connected to the photosphere by footpoints. Also the velocity field observed in the chromosphere is different from that observed in the photosphere (Ioshspa et al 1986, Martres et al 1981) and the inversion line of the chromospheric velocity is generally parallel to the axis of the filament.

The more recent observations of velocities in and around filaments are summarized in Table 2.1.

In filaments, the sign of the steady flow was in the past a controversial subject. Downflows were reported in prominences seen at the limb (Dunn 1960, Engvold 1976,Cui Shu et al 1985), whereas upflows were detected in filaments on the disk (Martres et al 1981, Malherbe et al 1983). The existence of a mean upward flow in the filament is , however, now generally accepted with an amplitude of 0.5kms^{-1} in H_α and of 5 kms^{-1} in CIV (Figure 2.4b,2.4d). A possible explanation for the inconsistency of this with limb observations could be that the latter represent fluctuations in the ionisation degree of material at different levels in the corona rather than a bulk motion.

Recent observations based on the analysis of H_α profiles obtained with the Multichannel Subtractive Double Pass (MSDP) spectrograph (Mein 1977, Schmieder et al 1984a,1985b, Simon et al 1986) and on spectrograms in the 10830Å HeI line (Figure 2.14 top, Engvold and Keil 1986) have confirmed the existence of a mean upflow in filaments.

Around filaments, the transition zone shows the same ascending behaviour as determined by several authors using CIV dopplergrams from data obtained with the Ultra Violet Spectrograph (UVSP) aboard SMM spacecraft (Athay et al 1983, Schmieder et al 1984a, Klimchuk 1986, Engvold et al 1985). In these studies the zero velocity obtained by averaging over the whole field of view (4'x4') is uncertain. Taking into account the general redshift expected in the transition region, the blueshifts around filaments would be reduced to 1.6 kms^{-1}, but recently the analysis of spectra obtained with HRTS aboard Spacelab 2 has shown velocities in CIV of the order of 5 - 6 kms^{-1} (Dere et al 1986).

An explanation for such a circulation of matter has been suggested by Malherbe and Priest (1983). They present 2D current sheet models supported against gravity by magnetic configurations with normal or inverse magnetic polarity compared with photospheric one, such as Kippenhahn-Schlüter (KS) or Kuperus-Raadu(KR). The new coronal plasma enters symetrically on both sides of the sheet where it condenses and cools. They suggest that upward motions in filaments could be due to photospheric motions causing a steady reconnection of magnetic field lines below a KR prominence. A correlation between these flows and large-scale convective motions in the photosphere (giant cells see section 2.1.3) has also been suggested by Schmieder et al (1984b). The upflow observed in filaments could balance the downflow observed generally around filaments in faculae. This scenario may also resolve the problem that one filament requires the total coronal mass to condense since the problem then becomes dynamical with plasma being carried up from the lower atmosphere (Tandberg-Hanssen 1986).

Moreover strong downflows comparable in H_α and CIV are observed (< 10kms^{-1}) at the ends or footpoints of filaments (Figure 2.4b). Their lifetime can be around 1 to 10 hours (Schmieder et al 1985b). Up and down motions ($\pm 6 kms^{-1} in H\alpha$) are also observed at footpoints suggesting the existence of loops (Figure 2.14, Kubota and Uesugi 1986) .

The situation becomes more confused for quiescent filaments observed with high resolution at the Pic du Midi (Figure 2.5b), even if the mean velocity is upward on the disk (Landman 1985a, Simon et al 1986, Démoulin et al 1987) where the velocity cell size is small and limited by the spatial resolution. The filament consists of vertical threads or fine loops imbedded in a large configuration and the velocity becomes enigmatic with no general behaviour; it is an average over many fine structures (Engvold and Keil 1986). The Dopplershift cells that we observe do not coincide spatially with intensity contours, so it appears that there is a problem due to the radiative transfer in the H_α line. The measured velocity and intensity probably do not come from the same region.

2.3.4 HORIZONTAL MOTIONS

Analysis of mass motions from centre to limb observations shows the existence of a horizontal flow in and around the filament. Fast horizontal motions (5kms^{-1}) slightly inclined to the prominence axis (20°)are suggested at the edges of active region filaments (Figure 2.15a, Malherbe et al 1983). The direction of the velocity is the same as that of the magnetic field lines.

This material circulation could involve a mechanism of stationary magnetic recon-

CHAPTER 2: OVERALL PROPERTIES AND STEADY FLOWS 35

(a)

(b)

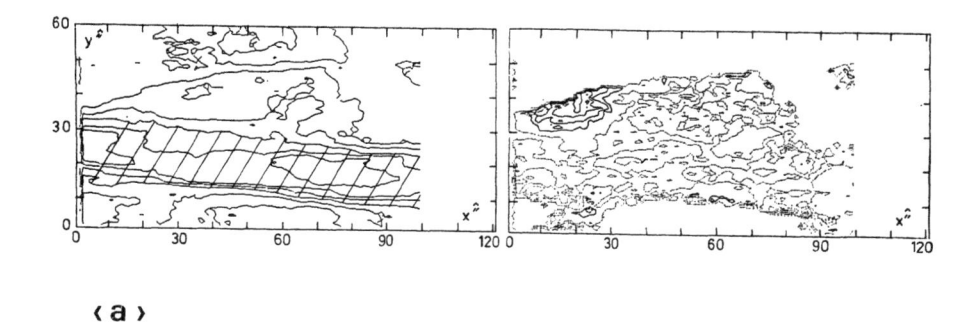

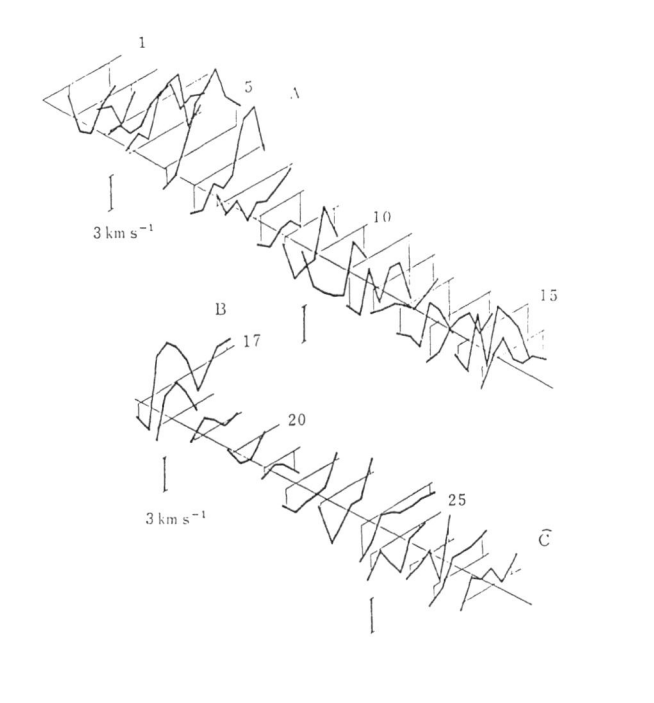

Figure 2.14: (a) Contour plots of a quiescent prominence (hatched area) observed in He 10830 Å at position N05 W22 (Engvold and Keil 1986). Intensity map (at the left), velocity map (at the right): blueshifts are in light lines, redshifts in solid lines. (b) The distribution of the vertical velocity along a dark filament. Vertical bars give a scale of 3 kms^{-1}. Positive velocities are downflows in this figure. A, B, C are the foootpoints (Kubota and Uesugi 1986).

nection using KR or KS support and a process of continuous material supply from the

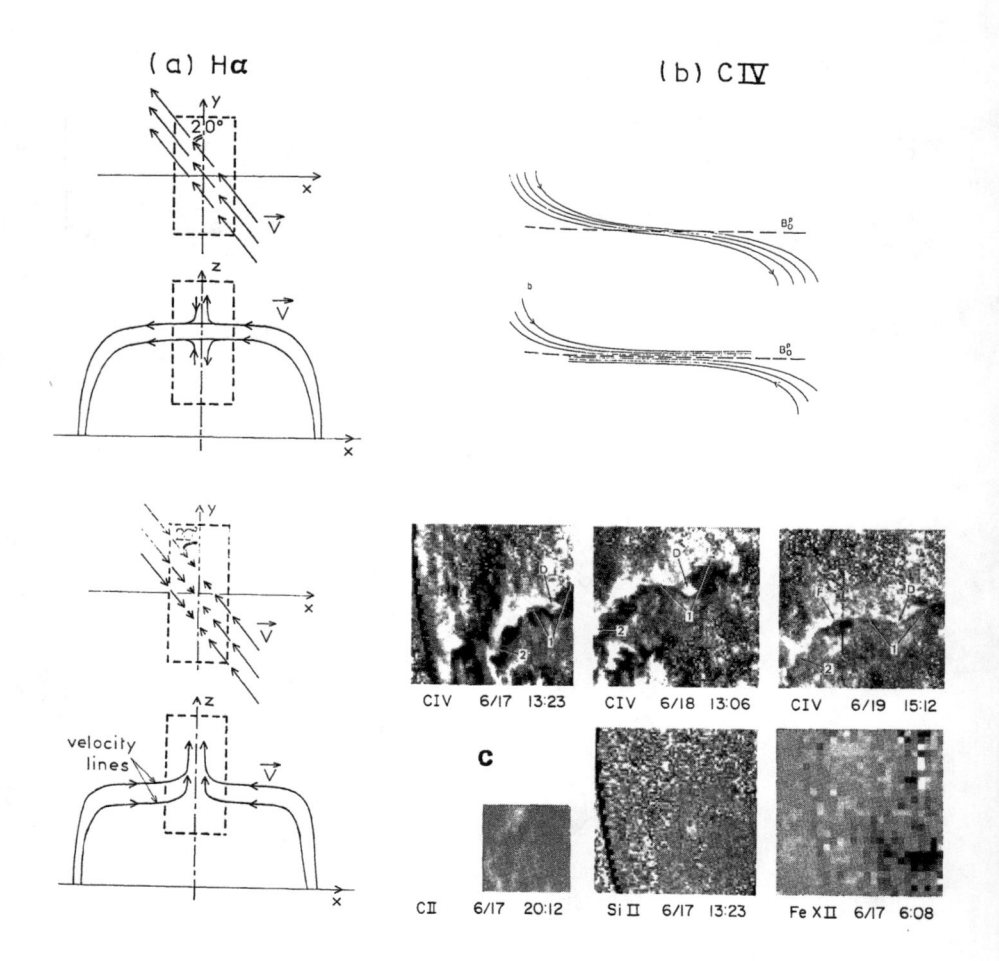

Figure 2.15: Schematic drawings of alternative magnetic field geometries deduced from dynamical behaviour (a) in the chromosphere (H_α) (Malherbe et al 1983) and (b) in the transition region (CIV) (Athay et al 1985) using velocity maps presented in (c).

chromosphere or the corona to the prominence. Such fast horizontal circulation has been reported by Vial et al (1979) in limb prominences observed in MgII and CaII lines and can be connected to observations at the active edges of prominences (Engvold et al 1978).

If we consider a model of magnetic loops where hot high-speed flows can decelerate and cool down to chromosphere values in order to form a filament (Ribes et Unno 1980), supersonic flows are expected in the transition zone. Horizontal velocities have been observed in EUV lines (Lites et al 1976, Vial et al 1980) and a sheared magnetic field in the transition region has been confirmed (Figure 2.15b, Athay et al 1985), but the velocity amplitudes are relatively small (< 20 kms^{-1}). The siphon model does not provide a consistent description of these observations which concern filaments in active regions. Simon et al (1986) observe a different behaviour in a quiescent filament in CIV, where the horizontal velocities are slightly less than the vertical velocities because of the integration through different structures looking like bushes of small loops aligned along the filament axis.

The problem of the mass flux may also be raised. The flux of cool H_α plasma should be equal to the flux crossing the transition region at the top of the magnetic flux tube. However, it is difficult to obtain a good accuracy for such measurements because of the uncertainty of the density values in H_α prominences (10^{16} - 10^{17}m^{-3}). The ratio of the velocity in H_α to that in CIV is found to be around 10 (Schmieder et al 1984a, Simon et al 1986) and so the conservation of the mass flux leads to a density ratio equal to 10. The density of the transition zone (TZ) that we can deduce is around 10^{15} to 10^{16} m^{-3} which is lower than measured by Poland and Tandberg-Hanssen (1983) around limb prominences (10^{17}cm^{-3}). This disagreement is due perhaps to the fact that the TZ observed on the disk results from the contribution of 2 regions, namely the TZ around the filament and the classical TZ below.

2.3.5 OSCILLATIONS

The problem of the existence of oscillations is well summarized in the review by Tsubaki (1988). Two kinds of oscillation have been detected in the velocity field , short period (< 10 min) and long period(> 10 min). Vrsnak (1984) and Wiehr et al (1984) detect oscillations between 3 to 8 min whereas Tsubaki and Takeuchi (1986), Balthazar et al(1986) and Tsubaki et al (1987) detect in spectrograms of prominences, observed on the limb, oscillation periods around 160, 210, 640 s (Figure 2.16). Such oscillations can be interpreted as waves (Alfvén) propagating along prominence loops (Solov'ev 1985).

Short period oscillations are, however, not detectable by other authors who observe at disk centre. Engvold (1981), for example, reported no clear evidence for oscillations or waves in simultaneous filtergrams taken in the two wings of the CaII line. Also, using the MSDP and the UVSP spectrographs, Malherbe et al (1981,1987) and Schmieder et al (1986) study radial velocities in H_α and in the CIV line and conclude that no oscillations are detectable (Figure 2.17).

They found an increase of the power spectra of the CIV line in the feet of a plage filament, but this could be due to waves or bulk flows. It would be useful to confirm the existence of short period oscillations in horizontal structures using Doppler-shift observations at different positions on the disk.

A study of long period oscillations has recently started. Bashkirtsev et al (1984, 1987) have been doing magnetographic observation in the H_α line and detect periods between 42 to 82 min. Wiehr et al (1984) and Balthazar et al (1986) using another method based on H_β Zeeman polarimeter observation find oscillations in the same range (48 - 64 min). Tsubaki (1988) concludes that periodic oscillations or waves can be amplified in prominences. However, the integration effect along the line of sight can blur the detection of periodicities over a wide range even if such phenomena exist, and the problem of fine structure certainly affects our study of this topic.

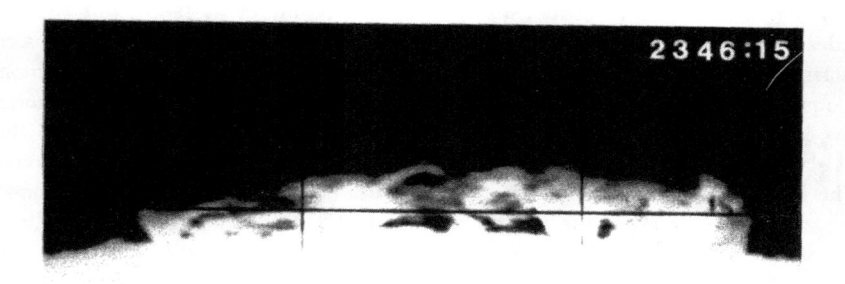

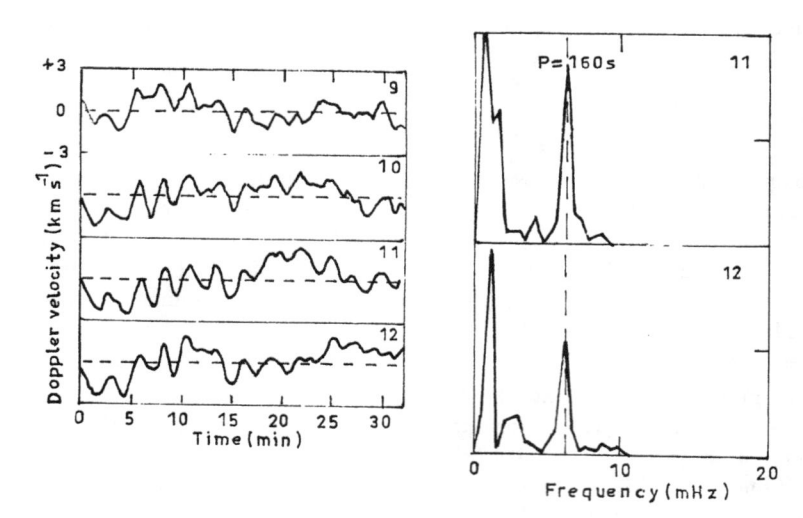

Figure 2.16: Oscillations with small periods detected in a H_α limb prominence (Tsubaki et al 1987)

2.4 Instability

2.4.1 DISPARITION BRUSQUE OF FILAMENTS

The phenomenon of sudden filament disappearance (DB) is a familiar one to observers in Hα (Figures 2.1 and 2.18, Martin 1980). Vizoso and Ballester (1987) reported that the number of DB per year is correlated with the number of activity centres, between 50 to 200 during the maxima and between 5 to 30 during the minima of activity. The North-South asymmetry curve of DB's (the difference number of DB's appearing in the northern and

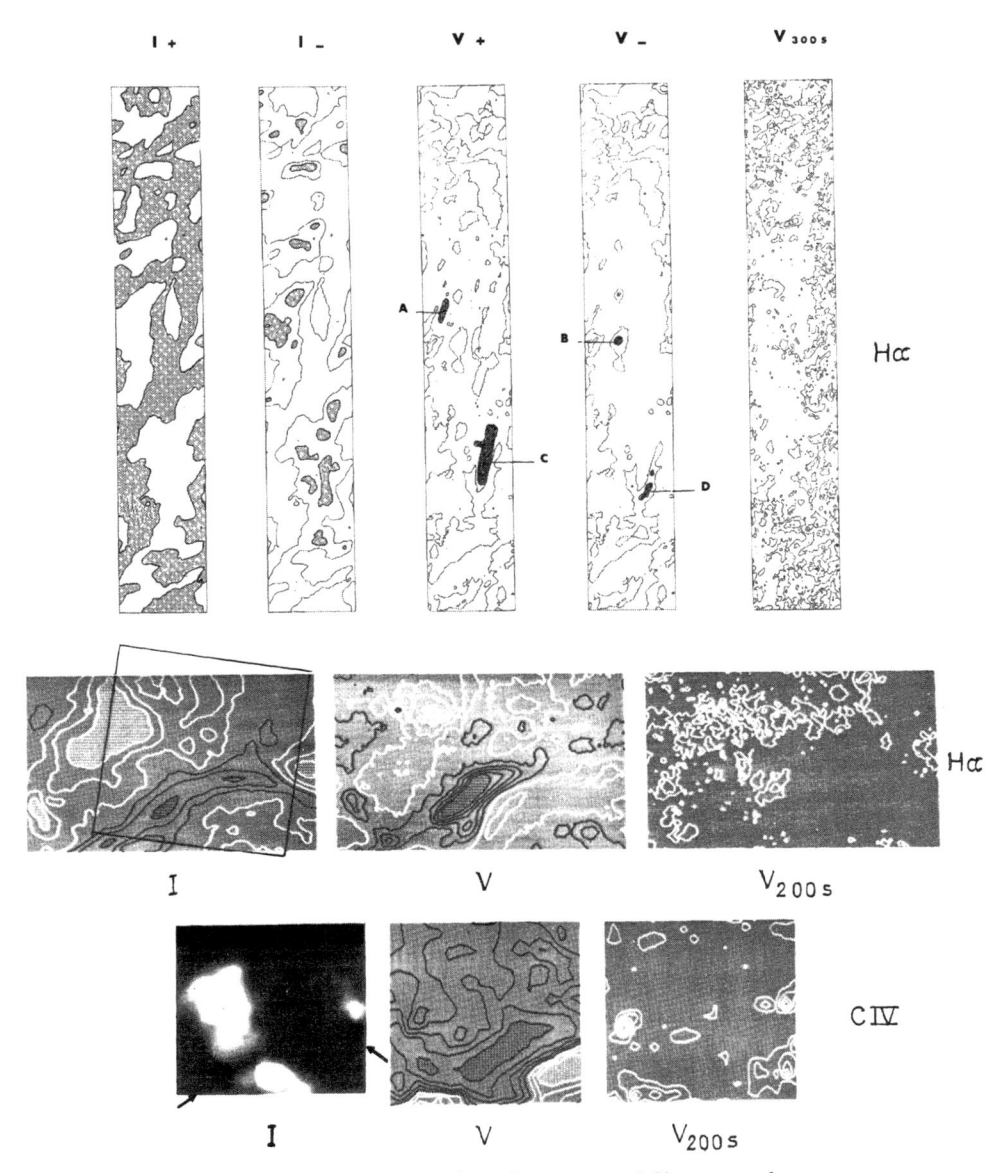

CHAPTER 2: OVERALL PROPERTIES AND STEADY FLOWS   39

Figure 2.17: 2D maps of I,V, energy at a given frequency of filaments observed in H_α and CIV on the disk (Malherbe et al 1981, 1987)

southern hemispheres normalized by the total number of DBs) is not in phase with the solar cycle but peaks at the minimum of the solar cycle, like the north-south asymmetry curve of flare numbers.

Hα - DB of a filament in June 1986

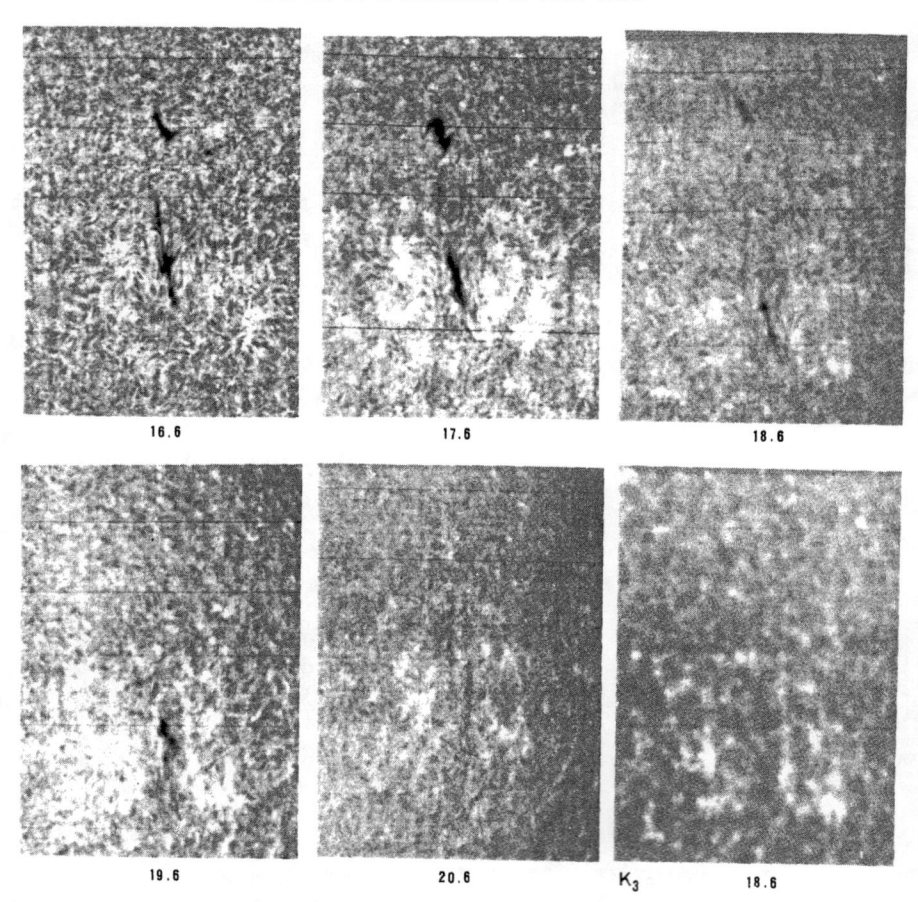

Figure 2.18: Disparition Brusque of a filament between June 15 to 20 1981. H_α intensity maps and a K_3 intensity map showing the supergranulation

A filament disappearance may be caused by heating or by dynamical processes similar in many respects to flare events (Rust 1984, Tang 1986). Disparition brusque of active region filaments precede the likely occurrence of flares (Kaastra 1985) but quiescent prominence DBs are not related generally to flares(Raadu et al 1987). DBs engender a growing interest because their study allows progress to be made in the understanding of filament formation and support(thermal stability) and of the causes and consequences of flaring instabilities with reconnection of magnetic field lines.

We can quote as an example of a thermal DB caused by heating, a Skylab observation of a filament (Figure 2.19). Mouradian et al (1986) show how it fades in $H\alpha$ and becomes

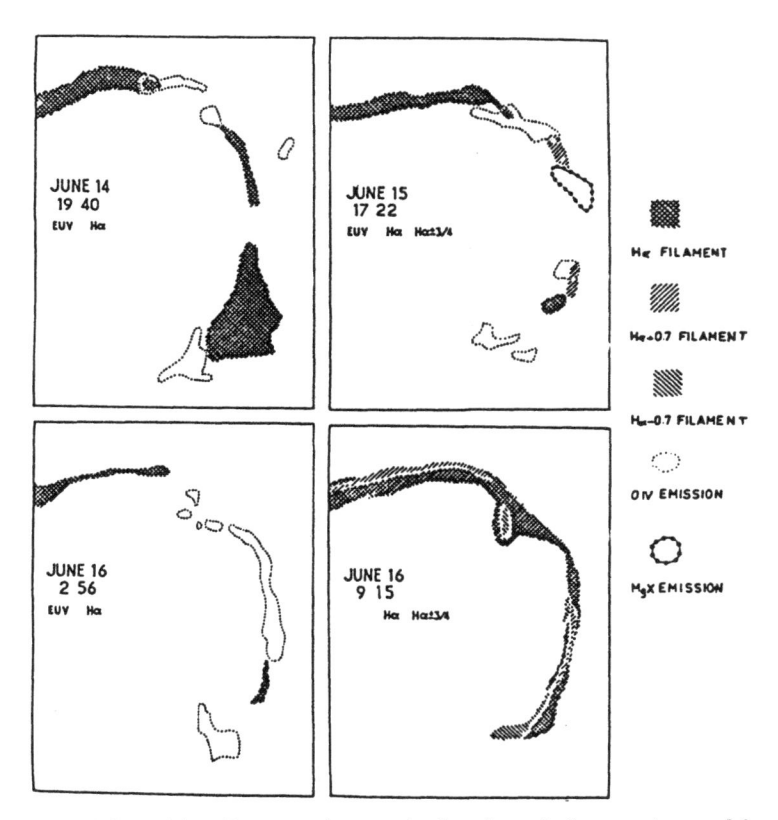

CHAPTER 2: OVERALL PROPERTIES AND STEADY FLOWS 41

Figure 2.19: Disparition Brusque due to the heating of plasma observed by ATM (Mouradian et al 1981).

visible successively in lines formed at increasing temperatures (CII, CIII, OVI, MgX). The loop observed by Hiei and Widing (1979) at 2.10^6K is probably a heated filament. Malherbe and Forbes (1986) investigate different mechanisms able to create such heating and conclude that strong magnetic heating or conduction such as in flares is necessary.

The trigger mechanism for disparition brusques is not well understood currently. Different mechanisms have been proposed including MHD non-equilibrium or instability (Chapter 7). Certainly magnetic forces of some kind are likely to be responsible for the acceleration of cool material in a filament. Raadu et al (1987) proposed a model where the acceleration is sustained by magnetic energy stored in the field of a moving plasmoid (Figure 2.20).

Observations of the destabilization of filaments (Figure 2.21) have been directly connected with photospheric activity such as the birth and shift of pores (Martin et al 1983, Simon et al 1984, Schmieder et al 1985a) or vortex motions (Martres et al 1982).

The relationship of photospheric activation and instability in filaments may be con-

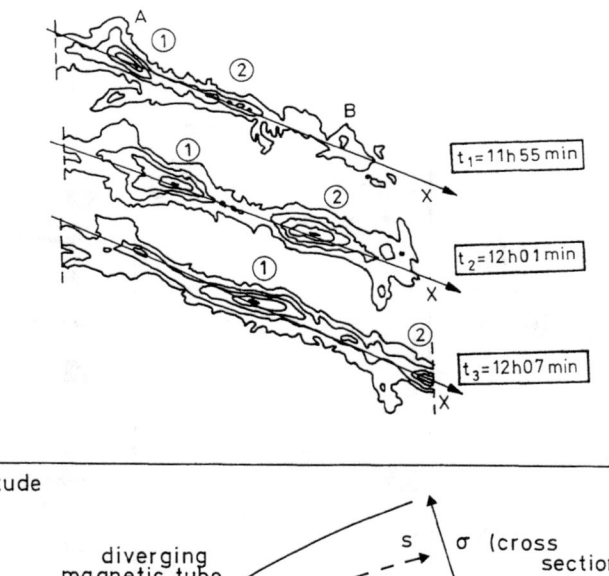

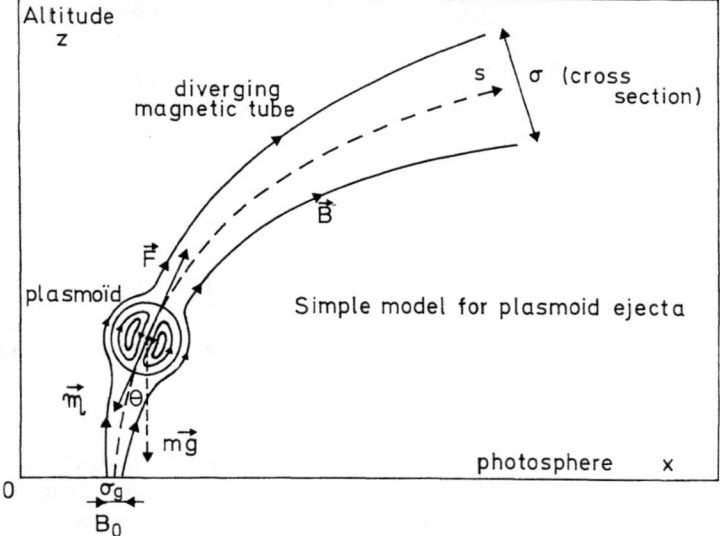

Figure 2.20: (a) Observations of a filament at three different times on June 22 1981 with the MSDP (contours of H_α intensity) and (b) schematic picture of the acceleration mechanism of magnetized plasmoïd in a filament interpreting the observations of the globules 1 and 2 in (a).

nected to the existence of pivot points (see section 2.1.3). Evidence has also been provided that the magnetic field configuration on a small scale such as new emerging flux (Simon et al 1986, Raadu et al 1988) or cancelling magnetic flux (Hermans and Martin 1986) can trigger filament activations. Footpoint motions and a sheared magnetic field may cause a filament to rise towards a new equilibrium (Kuin and Martens 1986). Hot loops visible in

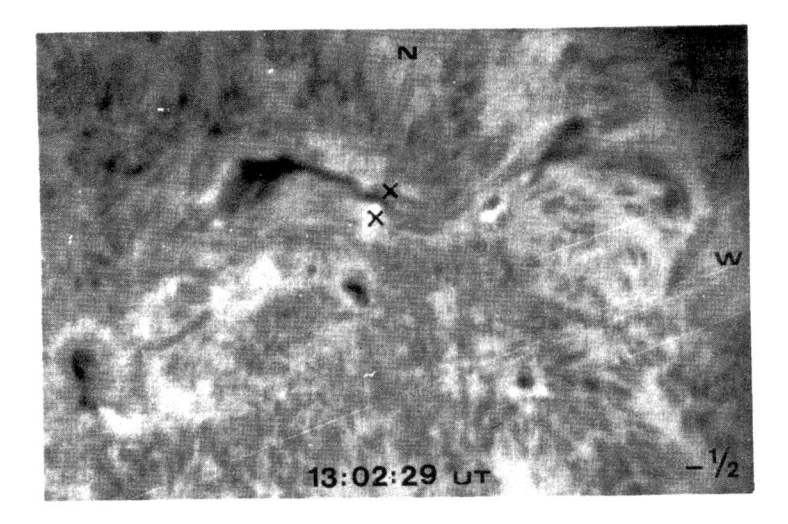

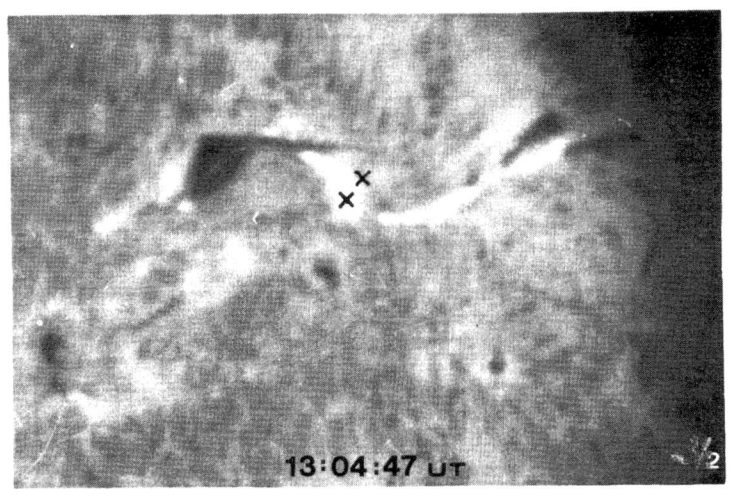

CHAPTER 2: OVERALL PROPERTIES AND STEADY FLOWS 43

Figure 2.21: Destabilization of a filament possibly due to the shift of a pore on June 22 1980 obtained at Debrecen Observatory (courtesy of L.Gesztelyi), the stars indicating the location of the pores.

X rays or in radio wavelengths, may also be destabilized filaments (Schmahl et al 1982, Gaizauskas 1985, Kundu et al 1985). The equilibrium of a filament may be due to a balance between upward magnetic buoyancy forces and the restraining forces of the overlying magnetic arches connecting regions of opposite polarity. If these restraining forces are too small the filament may rise, and erupt (Browning and Priest,1986).

Instabilities can propagate along the filament and lead to fast material ejecta (Figures 2.22a,2.22b, Raadu et al 1988) or to twisting or untwisting motions (Schmieder et al 1985a,

1988, Wang 1985, Kurokawa et al 1987). Liggett and Zirin (1984) have also studied proper motions of prominences and gave evidence of rotational motions.

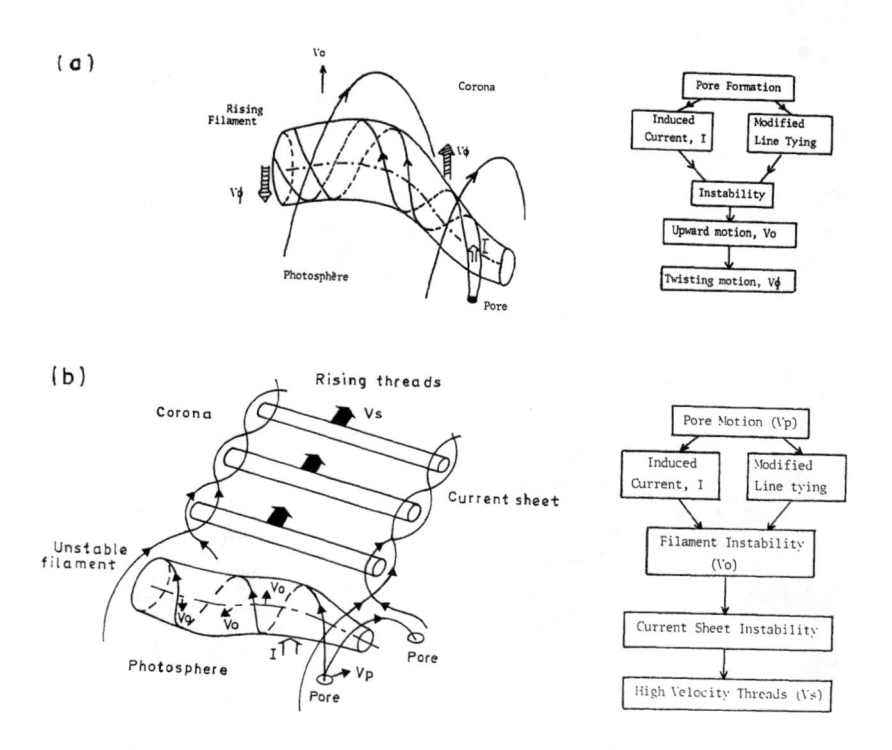

Figure 2.22: A schematic picture of destabilization of a filament by photospheric pores: (a) the filament rises and is twisted; (b) currents are created, material is condensed in a new current sheet, structured into threads by kink instabilities and accelerated by magnetic forces (Raadu et al 1988).

Helical motions were observed along a loop prominence by Gu et al (1984) (Figure 2.23) and by Prokakis and Tsiropoula (1987) . An early model was suggested by Malville (1978) and later Pneuman (1984) presented a model for rising helical prominences during activation. The dynamics of active-region prominences was also analysed by Klepikov and Platov (1985).

Another mechanism has been suggested by Apushkinskij and Topchilo (1984), relevant to deformations due to differential rotation. This can be related to singularities of rotation observed by Mouradian et al (1987) indicating new emergence of active centres. (see section 2.1.2)

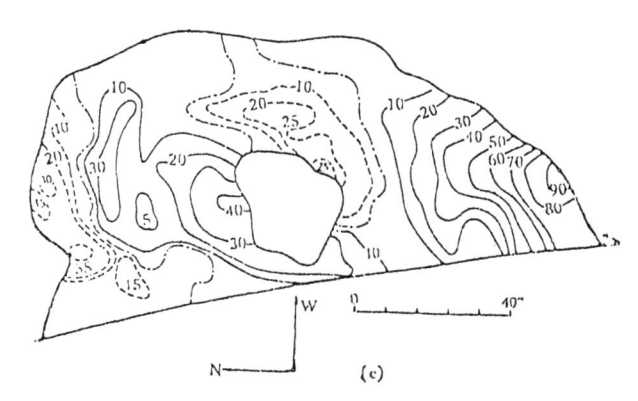

CHAPTER 2: OVERALL PROPERTIES AND STEADY FLOWS 45

Figure 2.23: Helical motions along a loop prominence observed on the limb. Solid lines refer to redshift velocity. Dashed lines refer to blueshift velocity. Dashed dotted lines show zero velocity. The unit of velocity is kms^{-1} (Gu et al 1984).

2.4.2 MODEL SUPPORT

From disappearing filaments we can learn about filament support. It is likely that the magnetic field lines are opened in the corona during a DB. Coronal mass ejections (Illing and Athay 1986, Athay and Illing 1986, Illing and Hundhausen 1986, Athay et al 1987) and particle acceleration (Kahler et al 1986) are often associated with a dynamical disappearance of filaments. DBs are known to be as important geophysically as flares for their geomagnetic storm, and ionospheric effects (Sastri 1987). The more common support is KR. Leroy points out that 75% of quiescent prominences have KR support and most of the prominences near active regions have KS support. It seems that the field around eruptive prominences is or becomes, in any case, KR-like. The filament reappears after the instability, which suggests that magnetic field reconnection occurs after destabilization (see chapter 4).

2.4.3 POST-FLARE LOOPS AND LOOP PROMINENCES

During the gradual phase of two-ribbon flares, dense and cold features called *post-flare loops* are formed . Reconnection models have been proposed by different authors (Kopp and Pneumann 1976, Forbes and Malherbe 1986)and are consistent with the observations. Post-flare loops show a large downward velocity along their legs reaching 60 kms^{-1}(Ishimoto and Kurokawa 1984, Hanakoa and Kurokawa 1986, Schmieder et al 1987). Heinzel and Karlicky (1987) computed theoretical redshifted profiles of $H\alpha$ lines to explain the brightening appearance of $H\alpha$ off-band observations of dense loops (Loughead et al 1983).

At the limb, bright loop prominences connected to flares are observed in $H\alpha$ with large velocities (Engvold et al 1979, Mein and Mein 1982). The mechanism accelerating cold material is more relevant to surge mechanisms than DB mechanisms, but we shall not discuss such phenomena here.

2.5 Conclusion

During recent years, great progress has been made in the understanding of filaments. We know the important role of filaments as tracers of large-scale structures and as indicators of flare onset. It is clear that filaments are dynamical events; global motions have been clearly established with large horizontal flows and slow upflows. Fine-scale structures of filaments have been resolved, and it has been shown how important it is to take these into account in thermal stability calculations.

Various parameters still remain uncertain, however, such as the density and ionization degree, which leads to a great uncertainty for mass flux conservation. A real problem is the understanding of the anchorage points of the filament in the photosphere : where are they located? how can the matter be stable vertically ? Observations of high resolution will be crucial, and measurements of the 3 components of the magnetic field are required. THEMIS (1990), the French telescope in the Canary Islands, will be invaluable here and the SOHO (1995) project should also determine the fine coronal structures around filaments.

CHAPTER 3

PROMINENCE ENVIRONMENT

ODDBJØRN ENGVOLD

Institute of Theoretical Astrophysics
University of Oslo
P.O. Box 1029, Blindern
N-0315 Oslo 3, Norway

Studies of prominences vis-a-vis the surrounding cornona provide insight into basic questions about promincence formation, their stability and support, mass and energy balance. The review discusses observational and theoretical results on helmet streamers, cavities, filament channels, and the prominence-corona transition sheath.

3.1 Introduction

The formation of quiescent prominences is inextricably tied to the questions of origin of their mass, the transport of energy between the prominence and the surrounding corona, and the topology of the supporting, and insulating, magnetic fields. It is assumed that the mass contained in a large quiescent prominence is a sizeable fraction of the total mass of the corona (Athay 1972; Tandberg-Hanssen 1974). Is the "cool" prominence plasma formed by condensation of hot

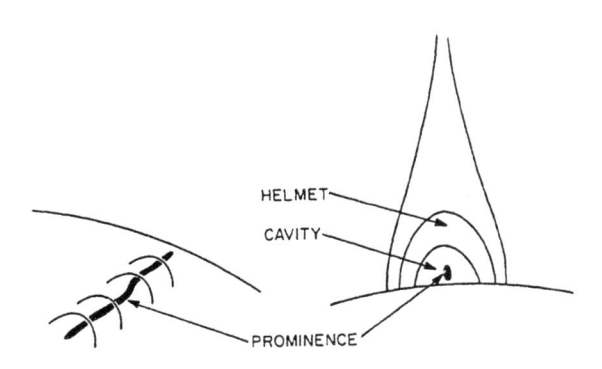

Figure 3.1. Topology of a typical prominence-corona helmet configuration. To the left it is seen on the disc whereas a limb view is seen to the right (Pneuman 1968).

coronal matter or by evaporation of chromospheric material? Several theoretical models have been put forward to explain the support of quiescent prominences, but the detailed mechanisms are not well understood (Chapter 6). Studies of the structure, dynamics, and thermodynamic conditions of the environment provide insight to the problems outlined above.

Quiescent prominences are located at the base of coronal helmet streamers. The central part of the helmet surrounding the prominence is less bright than the rest of the structure. This darker and less dense innermost volume is called a coronal cavity. Quiescent prominences are highly inhomogeneous. What actually belongs to the prominence and what is environment will be a matter of definition. The "cool" plasma is generally referred to as the prominence, and it is separated from its environment by a prominence-corona transition sheath. Figure 3.1 gives a schematic picture of the helmet streamer / coronal cavity / prominence complex. Flares are sometimes seen to preceed, and possibly trigger, prominence eruptions. Coronal Mass Ejections (CME's) are associated with erupting prominences. Actually, evidences seem strong that both limb flares and erupting limb prominences are preceeded by CME's (MacQueen 1988). These transient phenomena which also are rooted in the prominence environment are dealt with in Chapters 2, and 7.

3.2 Helmet Streamers

3.2.1. ECLIPSE PHOTOGRAPHY

Coronal streamers surrounding quiescent prominences have been seen in numerous eclipse photographs obtained since the beginning of this century. The coronal arch and streamer systems are faint and therefore difficult to record and measure accurately in the steep brightness gradient of the inner solar corona. Observers have supplemented their photographs by pencil drawings to enhance the fine structure of the helmet streamer systems. The beautiful color photographs of Figure 3.2 taken at the 1983 total solar eclipse in Java (Tsubaki 1986) illustrate the clear association between large quiescent prominences and coronal streamer structure, but they also show cases where no connection is obvious. The observational technique has been improved in recent years by using radially graded neutral density filters in the telescope focus resulting in an enhancement of the wealth of fine details of the structure. Saito and Tandberg-Hanssen (1973) reviewed eclipse observations of 36 different helmet streamers taken in the period 1896–1970, and they presented a detailed analysis of a well observed large helmet streamer system observed at the eclipse in Peru, November 12, 1966.

3.2.2 MORPHOLOGY

In the helmet streamer of the Peruvian eclipse studied by Saito and Tandberg-Hanssen (1973) one noticed that at least two arch systems are encapsuled by the streamer (see Figure 3.3). Similar twin-arch streamers are seen in earlier photographs (Eddy 1973). One generally finds that individual streamers are not associated with single active regions or other particular disk features (Bohlin 1970; Poland 1978; Poland and MacQueen 1981). Eddy (1973) showed from his studies of eclipse photographs that individual magnetic field lines of the helmet can be formed by the

CHAPTER 3: PROMINENCE ENVIRONMENT 49

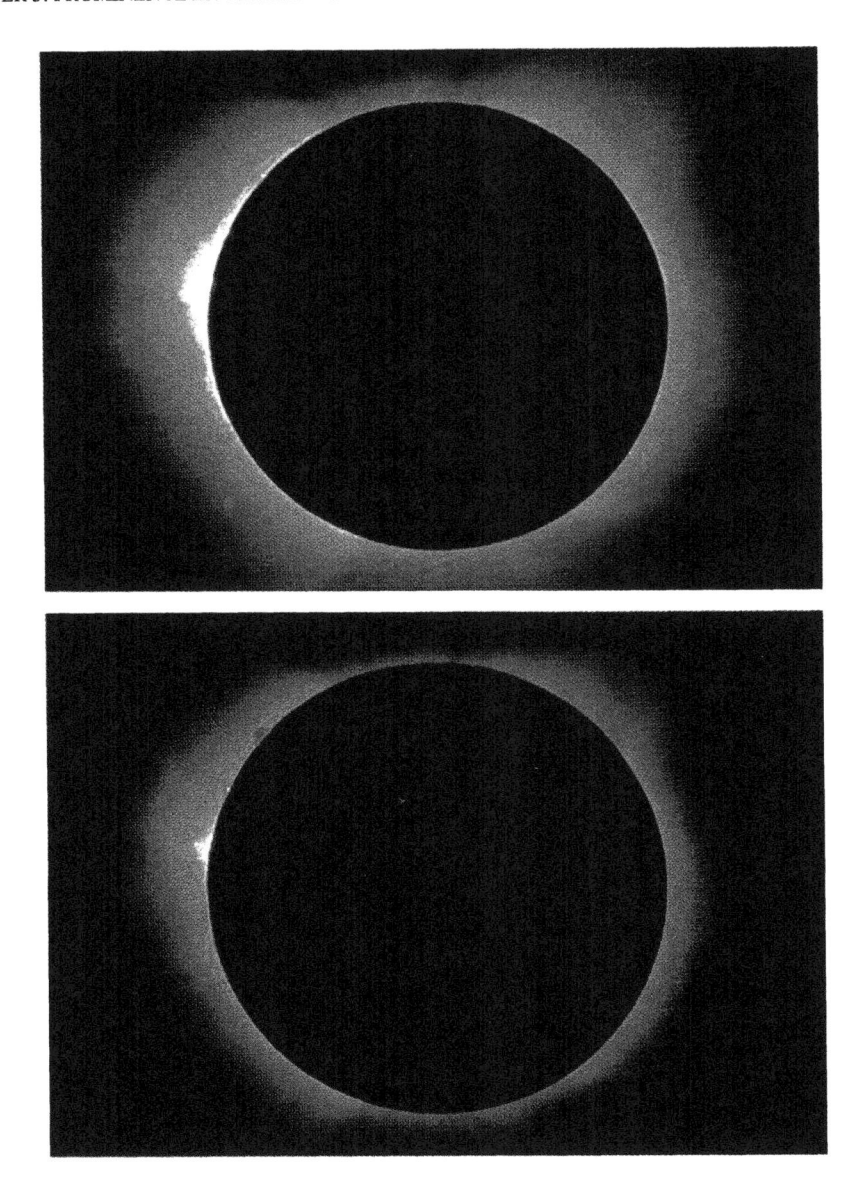

Figure 3.2. Coloured images of the innermost corona photographed during the total eclipse over Java, June 11, 1983. The prominences are red because of their strong emission in the Hα line at λ6563Å. The upper picture was taken at $11^h\ 30^m 14^s$ with an exposure time of 1/30 s, and the lower taken 7s later with 1/60s (Courtesy T. Tsubaki).

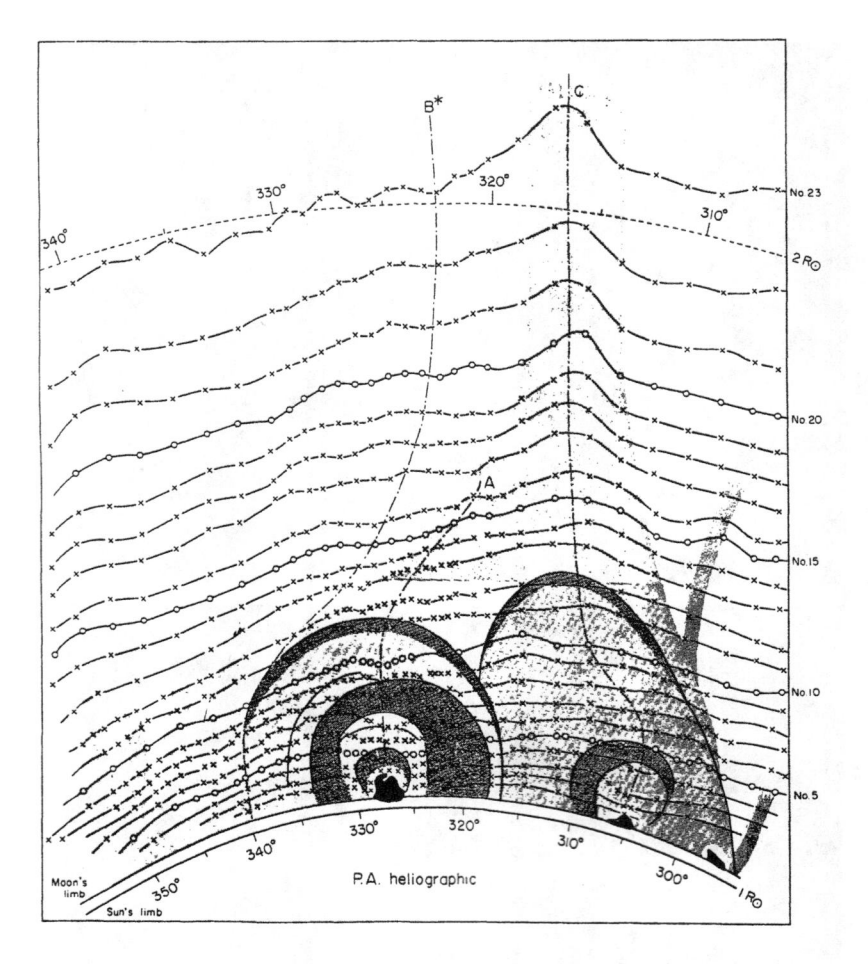

Figure 3.3. Brightness map of a large helmet streamer observed during the Peruvian eclipse of November 12, 1966. A and C trace the centre lines of the two arch systems (Saito and Tandberg-Hanssen 1973).

confluence of open field lines form two widely separated bipolar magnetic regions which each forms a system of high magnetic arches. This picture is supported by the study of McCabe (1986) who found that the white-light emission detected with the K-coronameter is cospatial with low-resolution (180 arcsecond^2) large-scale images of the photospheric neutral lines.

Coronal streamers are detected out to distances from the Sun's limb of more than one solar radius. Waldmeier (1972) measured cross sections of typically 30 000 km for long thread-like streamers in the solar corona. Helmet streamers are often found to extend over ranges of 50 degrees, and more, in heliocentric latitude. A typical diameter for a helmet is 300 000 km and its projection on the surface exceeds 500 000 km in the major dimension. The heights of the closed

CHAPTER 3: PROMINENCE ENVIRONMENT   51

loop helmet structure seen in eclipse photographs is comparable to its diameter (Kawaguchi 1967; Saito and Tandberg-Hanssen 1973).

Davis and Krieger (1982) detected hot coronal loops striding over quiescent prominences. Their footpoints are found near supergranulation cell boundaries.

3.2.3 LOCATION OF CURRENT SHEET

The conceptually simple picture of corona streamers is that they are associated with, and overlie, filament channels, i.e. regions of reversed magnetic polarity in the photosphere, and possibly some active regions, and as such they must contain a neutral sheet. The presence of a neutral

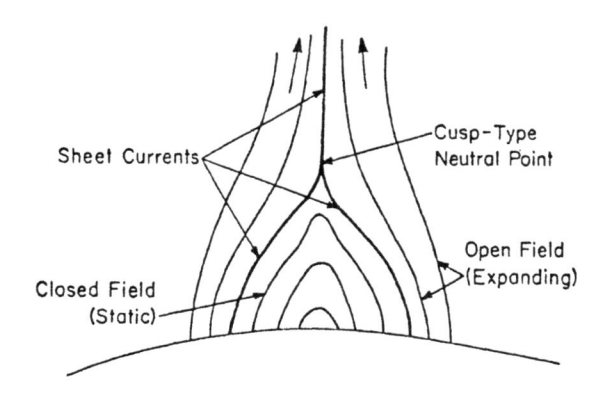

Figure 3.4. Schematic diagram of a typical helmet streamer configuration (Pneuman and Kopp 1971).

current sheet is not obvious from white-light photographs. A coronal current sheet may be be too thin to be resolved by conventional eclipse instruments and K-coronameters. Sturrock and Smith (1968) identified a feature as a possible current sheet in an eclipse photograph taken in 1965. A possible neutral point of inverted Y-shape was noticed at the top of a low coronal arch. Figure 3.4 shows a schematic diagram of a typical helmet streamer configuration.

3.2.4 BRIGHTNESS

It is found that the streamer brightness bears no simple relation to the magnetic field strenght at the base (Frazier 1972). The field geometry. rather than the strength, seems to be the underlying factor that determines the formation and brightness of a streamer (Pneuman 1973).

3.3 Coronal Cavities

3.3.1 BRIGHTNESS AND STRUCTURE

It is well known from eclipse photographs that quiescent prominences are imbedded in regions of the corona that are darker than the average of the associated helmet streamer (Leroy and Servajan 1966; Kawaguchi 1967; Saito and Tandberg-Hanssen 1973). Some quiescent prominences do not seem to have an overlying streamer, but this may possibly be caused by too short an exposure and because some streamers are very faint. The reduced coronal brightness of a cavity region is illustrated by Figure 3.5.

Some coronal cavities contain concentric multiple arch systems (Saito and Hyder 1968). Koutchmy (1987a) finds pronounced structures in white-light and Fe XIV 5303Å images of coronal cavities. The height and width extension of the cavity region is roughly twice as large as the prominence (see Waldmeier 1970; Saito and Tandberg-Hanssen 1973).

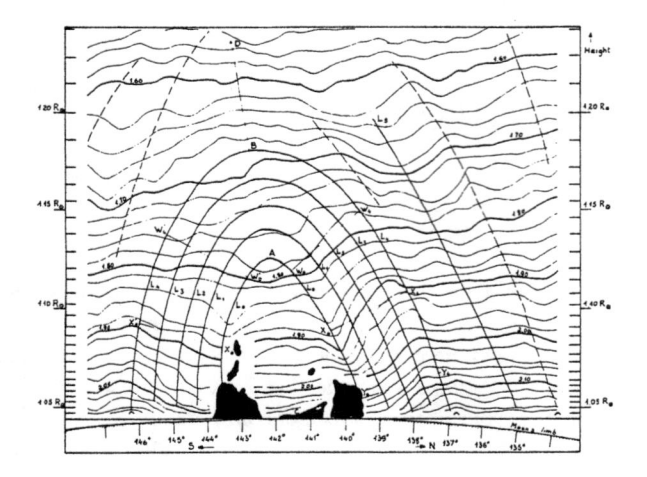

Figure 3.5. Isophotes of the dome formation around a prominence. The numbers attached to the isophotic curves give intensities of the K-corona in units of 10^{-8} times the mean brightness of the solar disc (Kawaguchi 1967).

CHAPTER 3: PROMINENCE ENVIRONMENT 53

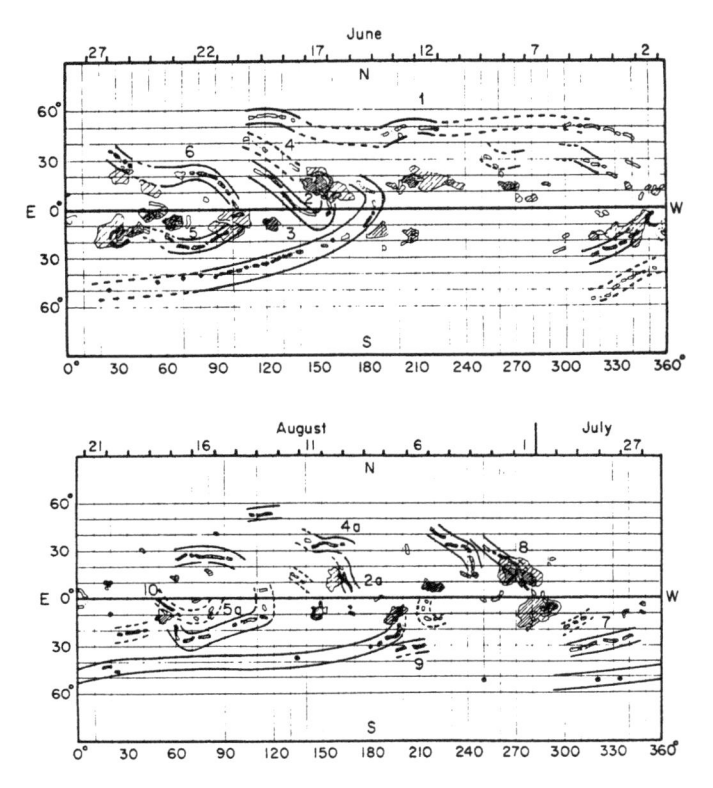

Figure 3.6. Outlines of X-ray cavities identified in images recorded with the S-054 instrument on Skylab. The associated filaments are taken from the Meudon Hα synoptic charts (Serio et al. 1978).

X-Ray images that were taken with the S-054 instrument on board Skylab revealed regions of low emission. These were called coronal X-Ray cavities (XRC) as a result of their spatial correlation with prominences. Figure 3.6. gives the outlines of the X-ray cavities detected by Serio et al. (1978) super-imposed on Hα synoptic charts. The authors found coronal cavity widths of 60 000 km and height > 50 000 km. Cavities are evident also in the radiation of EUV lines formed at coronal temperatures, such as lines of Ne VII and Mg X (Noyes et al. 1972; Schmahl et al. 1974). However, in transition region lines, like Si IV and C IV formed around 100 000 K, the contrast is very low and cavity regions are hardly detectable.

3.3.2. TEMPERATURE AND DENSITY

A coronal cavity is a region of reduced electron density (Waldmeier 1941, 1970). On the other hand, the coronal temperature is found to be the same inside and outside the cavity (Bessey and Liebenberg 1984).

3.4 Filament Channels

3.4.1 ASSOCIATION WITH NEUTRAL LINES

It has long been known that prominences always overlie so-called neutral lines, i.e. the polarity reversals of the photospheric longitudinal magnetic fields (Babcock and Babcock 1955; Smith and Ramsey 1967; McIntosh 1972). The neutral lines dividing the large-scale regions of magnetic polarity will by necessity be continuous.

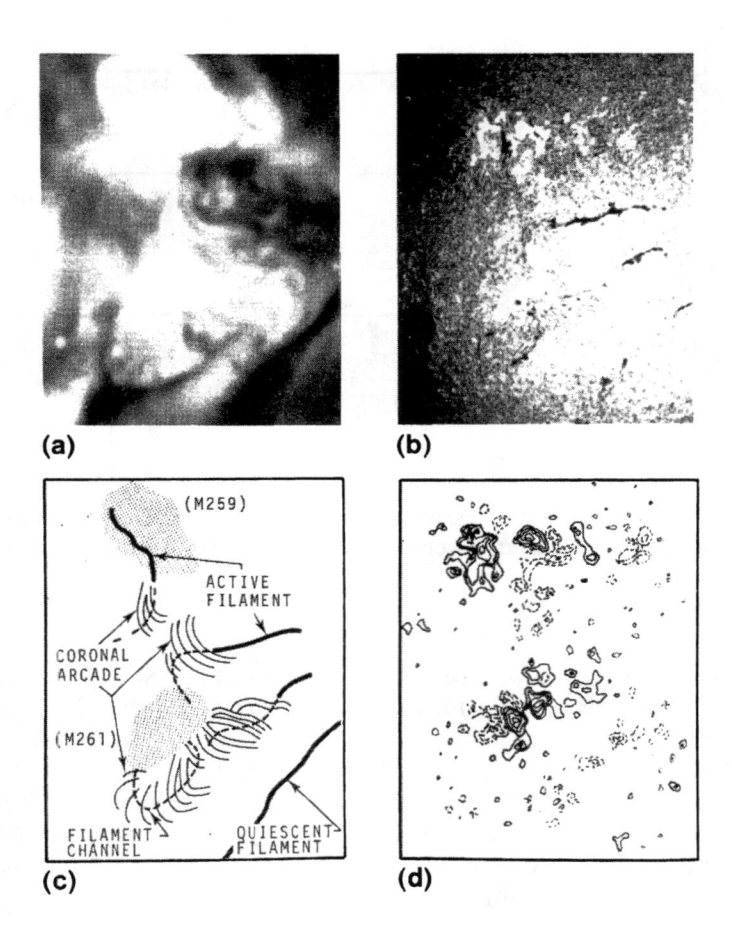

Figure 3.7. The association of X-ray, Hα and magnetic features. (a) X-ray image showing the arcades of bright loops and coronal cavities. (b) Simultaneous Hα photograph showing the location of the filaments and filament channels. (c) A schematic showing the location of coronal arcades above active filament channels. (d) The photospheric magnetic field obtained at Mt. Wilson (Davis and Krieger 1982).

CHAPTER 3: PROMINENCE ENVIRONMENT

Nearly all X-Ray structures overlying a neutral line are in the form of arches (Figure 3.7), resulting in the helmet structure when seen at the Sun's limb. One may picture the magnetic field in these regions as twisted up along the neutral line to form a continuous structure of the shape of a cocoon, everywhere shielding the prominence below from the surrounding hot corona.

High-resolution magnetograms do clearly show that opposite polarity fields exist on either side of the polarity reversal line (Figure 3.8; Martres et al. 1981). In fact, the formation of dark filaments seems to follow cancellation of small fragments of opposite polarity close to the neutral line (Martin 1986). Whenever a bipolar element exists close to the neutral line, one finds a break of some sort in the prominence structure (Zirin 1988).

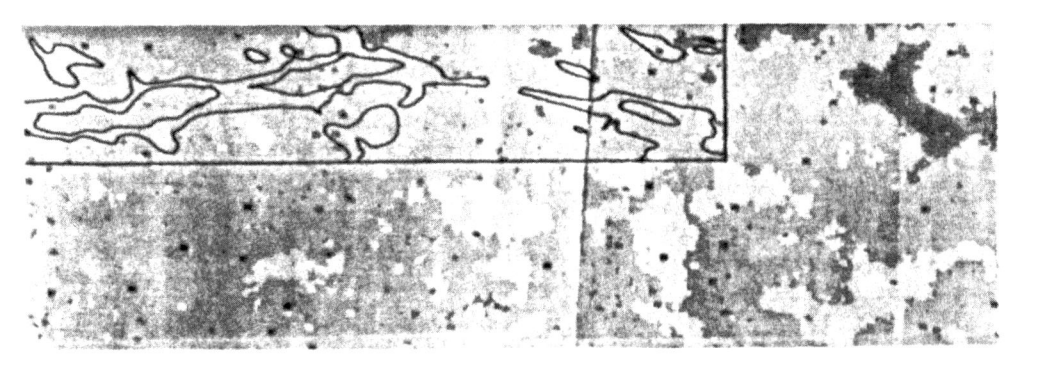

Figure 3.8. A map of the magnetic field observed with the MSDP inside and near a filament channel (11 October 1977). An isophotic Hα image shows the position of the filament channel. Note the many small bipolar magnetic fields in the channel (Martres et al. 1981).

3.4.2 POLEWARD MIGRATION OF FILAMENT CHANNELS

High latitude filament channels are oriented chiefly east-west form nearly uninterrupted circumpolar crowns. Observations show that meriodinal flows carry the magnetic regions and the associated filament channels polewards (Topka et al. 1982). The average speed of the poleward drift ranges between 6 and 12 m s^{-1}.

3.4.3 PRESENCE OF PROMINENCES

A prominence usually appears morphologically as several disconnected elements congregated along the filament channel (Figures 3.6 and 3.9). The fraction of the total cavity length in which prominences are observed in Hα varies from 0.1 to 0.6. The actual value seems to depend on the

Figure 3.9. Full disk Hα filtergram of September 20, 1981 recorded at Norikura Solar Observatory (Courtesy H. Morishita).

age of the corresponding neutral line. Older filament channels contain smaller and short-lived elements of prominences (Figure 3.10). The character and strength of a prominence (filament) is not the same in different lines. A comparison of the same filament seen in Ca II K and H I Lyα show very similar structures, whereas the filament was noticeably different in Hα observations of Bonnet et al. (1980) (see their Plates L3 and L4). See also Figure 3.11. The difference in appearance is evidently caused by different radiative transfer mechanisms in the lines.

CHAPTER 3: PROMINENCE ENVIRONMENT

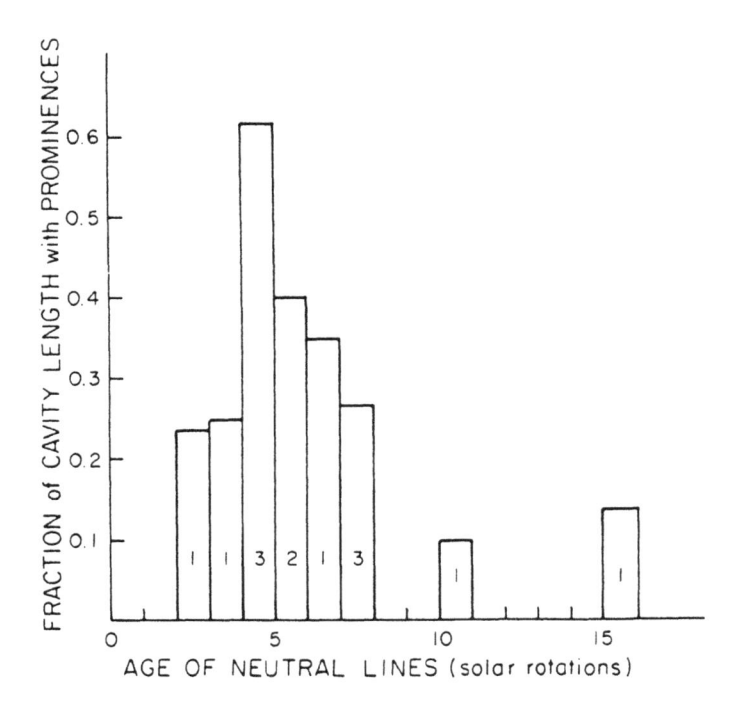

Figure 3.10. Histogram showing the fraction of the total cavity length in which prominences (filaments) are present as a function of the age of the corresponding neutral lines in units of solar rotations (Serio et al. 1978).

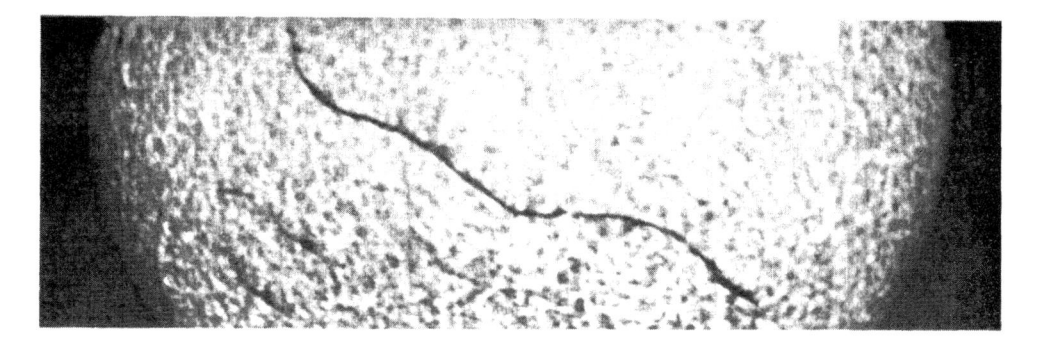

Figure 3.11. Ca II K spectroheliogram of a large filament observed on 20 July, 1922. Notice the continuous appearance of the dark disk filament (d'Azambuja and d'Azambuja 1948).

1984 JULY 17 (36 GHz)

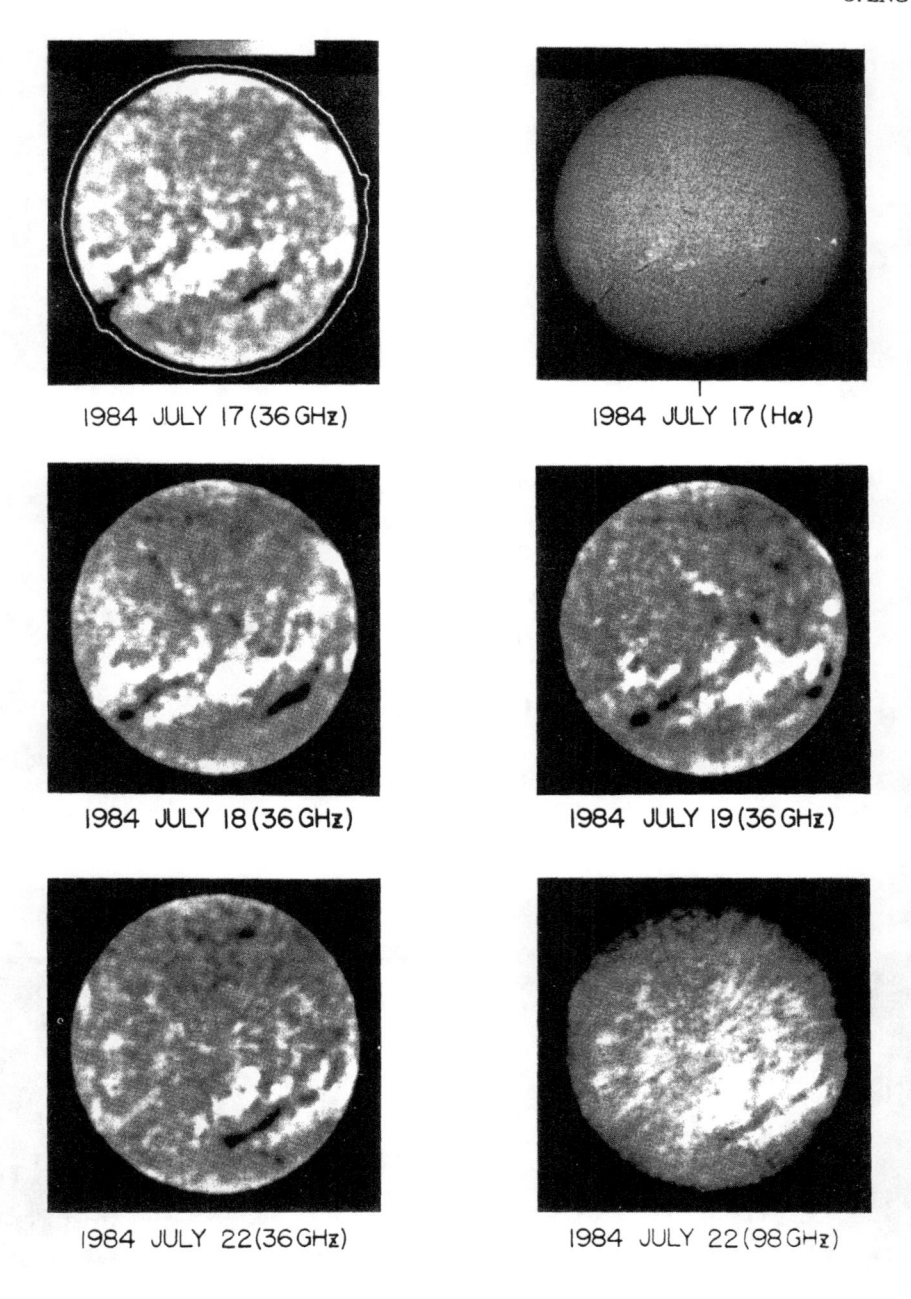

1984 JULY 17 (Hα)

1984 JULY 18 (36 GHz)

1984 JULY 19 (36 GHz)

1984 JULY 22 (36 GHz)

1984 JULY 22 (98 GHz)

Figure 3.12. Radio filaments observed at 36 GHz (6 mm) and 98 GHz (3.1 mm). The locations of dark Hα filaments are seen in the upper right hand image (Courtesy E. Hiei).

CHAPTER 3: PROMINENCE ENVIRONMENT 59

Dark filaments are also detected as regions of reduced emission at radio wavelengths. Observations at 6 cm and 20 cm wavelength (Kundu 1986), and 3.1 mm and 8.3 mm (Hiei et al. 1986) show radio emission voids ("radio filaments") which are cospatial with, but larger than their optical counterparts (Figure 3.12). One always finds a radio filament with every Hα filament, but many radio filaments have no observable Hα filament. The noted variation in darkness and filling of the filament channel from one line, or type of emission, to another may mean that there is more cool material than we actually see in Hα. A detailed analysis of such observations is needed.

High-resolution (2 arcsec) observations of quiescent prominences with the VLA have revealed the presence of double ribbons of enhanced microwave (15 GHz) emission along the prominences (Gary 1986). The emission bands are about 7000 km wide, which seems too wide to be produced by the P-C transition sheath. The brightness may rather be associated with the foot-points of the local magnetic fields that serve to support the prominences (see also Section 3.6.1).

Studying EUV ionic lines formed at temperatures up to 2 million K (Fe XV), Chapman (1981) claimed that there exist two types of filaments (prominences) with respect to observed brightness contrast. Chapman points out that the brighter filaments seem to occur in regions with large surface gradients in the magnetic field.

3.4.4 TEMPERATURE AND ELECTRON PRESSURE

The electron pressure of the prominence-corona (P-C) interface region, which can be measured from the intensity ratio of density sensitive line pairs, is nearly the same as in the cavity. Filament channels are recognizeable through their lower brightness in various wavelength regions regardless of whether there is a prominence or not. In X-Ray cavities the presence of a prominence segment seems to influence the cavity brightness such that it is slightly lower than when no prominence is seen (Serio et al. 1978). The X-Ray cavity (and EUV cavity) region without observable prominence material yields an electron density $3-4 \times 10^8$ cm^{-3}.

3.4.5 COOL MATTER IN THE FILAMENT CHANNELS

The presence of "cool" absorbing matter in the filament channels is more pronounced at radio frequencies than in Hα. The difference is most likely an effect of opacity. The cool material may exist in too small amounts to be observable in lines such as Hα, but yet be enough to be noticeable at mirco-wave frequencies. In the case of electron density $N = 3 \times 10^{10}$ cm^{-3} an optical depth unity ($\tau \simeq 1$) corresponds to 40 km at 36 GHz and 280 km at 97 GHz (Hiei et al. 1986). The corresponding length in Hα is larger and about 1000 km for $N_e \simeq 4 \times 10^{10}$ cm^{-3} and T = 6000–7500K according to data from Engvold (1976). Two alternative scenarios are apparent; either (1) the cool structures are very thin and few, or (2) the gas density is low ($N_e < 10^{10}$ cm^{-3}).

The sum of the magnetic pressure and the gas pressure in the cavity region balance the ambient pressure in the corona. The local gas pressure may therefore be less than the coronal pressure. Local fluctuations in the magnetic pressure may lead to inverse variations in the gas pressure. Koutchmy (1987b) has pointed out that the inhomogeneities in the lower corona represent variations in density (N_e) by as much as two orders of magnitude.

The equilibrium condition in the prominence gas is dictated largely by the photo-ionization rate from the ground level in the hydrogen atoms and the subsequent electron recombination. Assuming a pure hydrogen gas, the ionization balance may be expressed by:

$$\alpha_{c1} N_e^2 - R_{1c} N_H \simeq 0 \tag{3.1}$$

where N_e and N_H are the number density of electrons and hydrogen atoms, α_{c1} is the coefficient of recombination and R_{1c} is the photoionization rate per atom. One may express the relation in terms of fractional ionization, i.e. $X = 1-(N_H/N)$ where N is the total number density of hydrogen particles ($0 \le X \le 1$). Since $N_e \simeq XN$ and $N_H = (1-X)N$ eq. (3.1) becomes

$$X^2 \alpha_{c1} N - (1-X) R_{1c} \simeq 0 \tag{3.2}$$

Any value of N, i.e. of the gas pressure (P = Nkt), will result in an equilibrium value of X according to Equ. (3.2).

Using $\alpha_{c1} \simeq 2 \times 10^{-13}$ (Allen 1973) for typical prominence temperatures, and $R_{1c} \le 3.6 \, \alpha \, 10^{-3}$ s^{-1} (see Engvold 1980), one may calculate the variation of N_e and N_H with gas pressure P. The result is given in Figure 3.13. The Hα absorption is proportional to the population of the secondary energy level of the hydrogen atoms, N_2, which relates to N_e according to (cf. Engvold 1976):

$$N_e \simeq 3.3 \times 10^8 \sqrt{N_2} \tag{3.3}$$

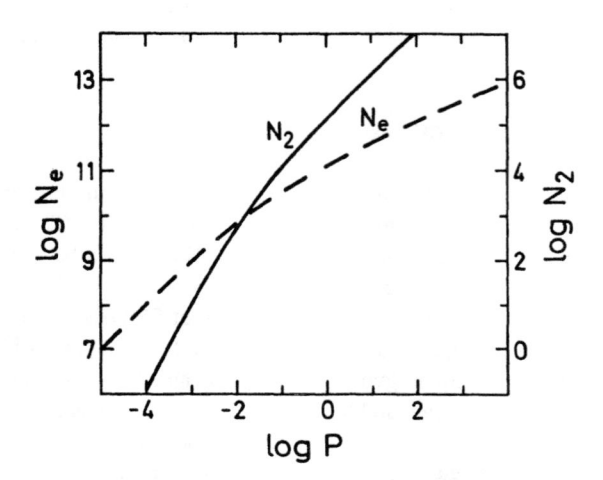

Figure 3.13. Variation of N_e (cm^{-3}) and the population of the secondary energy level of hydrogen atoms (N_2) with pressure in a pure hydrogen gas at prominence temperature T = 7500 K (see text). The gas is assumed to be optically thin for the incident, ionizing, Lyman continuum radiation from the chromosphere.

By combination of Equ. (3.1) and (3.3) one finds $N_2/H_H \simeq 1.7 \times 10^{-7}$. For $N < 3 \times 10^{10}$ cm^{-3} ($P < 0.03$ dyn cm^{-2}) the number of neutral hydrogen atoms, and hence the population of the second level, decreases more rapidly than N (and N_e). In other words, as the gas pressure in the cavity region decreases relative to the ambient coronal pressure the hydrogen gas becomes rapidly transparent to Balmer Hα radiation. The gas may, however, still be opaque at radio frequencies.

3.5 Prominence-Corona Transition Region

3.5.1 LINE EMISSION FROM THE TRANSITION REGION

Analysis of EUV and UV lines in quiescent prominences show that the transition region between prominences and the corona is very thin and in general similar to the chromosphere-corona transition region (Yang et al. 1975; Orrall and Schmahl 1976; Schmahl 1980, 1986; Rabin 1986).

New observations by Noëns (1987), which record simultaneously the high temperature coronal lines (Fe xiii 10747Å and 10798Å) and the low temperature line He I 10830Å, indicate that there is an enhancement in electron density close to the P-C interface (Figure 3.14). In the case of constant pressure through the transition region one should in fact expect the density to increase with decreasing temperature. Smartt and Zhang (1984) detected a similar prominence edge enhancement in the coronal lines Fe x 6374Å and Fe xiv 5303Å. The red- and green-line images were interpreted as representing the P-C transition region. The authors noted that I(6374)/I(5303) $\simeq 2$ which was different from and actually reversed compared with the line emission typical for other coronal structures. Since these two lines are formed at somewhat different coronal temperatures their ratio contains information about the temperature structure of the P-C region.

Analysis of UV emission lines formed in the P-C transition sheath give electron pressures between 1/2 to 1/5 times the pressure in the normal transition region and corona (Noyes et al. 1972; Orrall and Schmahl 1976; Kjeldseth-Moe et al. 1979; Schmahl 1979; Mariska et al. 1979).

3.5.2 EMPIRICAL MODELLING OF THE P-C TRANSITION REGION

The empirical modelling of the transition region begins with a determination of the emission measure from UV and EUV lines. The method has been elaborated by a number of investigators (Athay 1966; Dupree 1972; Jordan and Wilson 1971; Raymond and Doyle 1981; Schmahl and Orrall 1986). Following Nicolas et al. (1982) and Kjeldseth-Moe et al. (1984) the differential emission measure may be expressed by:

$$F(T) = A_{eff} P^2/(dT/dh). \qquad (3.4)$$

Here A_{eff} is the effective radiating area ("filling factor"). P_e is the electron pressure, and dT/dh the temperature gradient along the line of sight. In this case the ralation between F(T) and intensity for resonance lines is written:

$$I_v = 2.65 \times 10^{16} A_{EL} <g>f \int F(T) R_i(T) T^{-3/2} \exp(-h\nu/kT) d(\log T), \qquad (3.5)$$

Hα

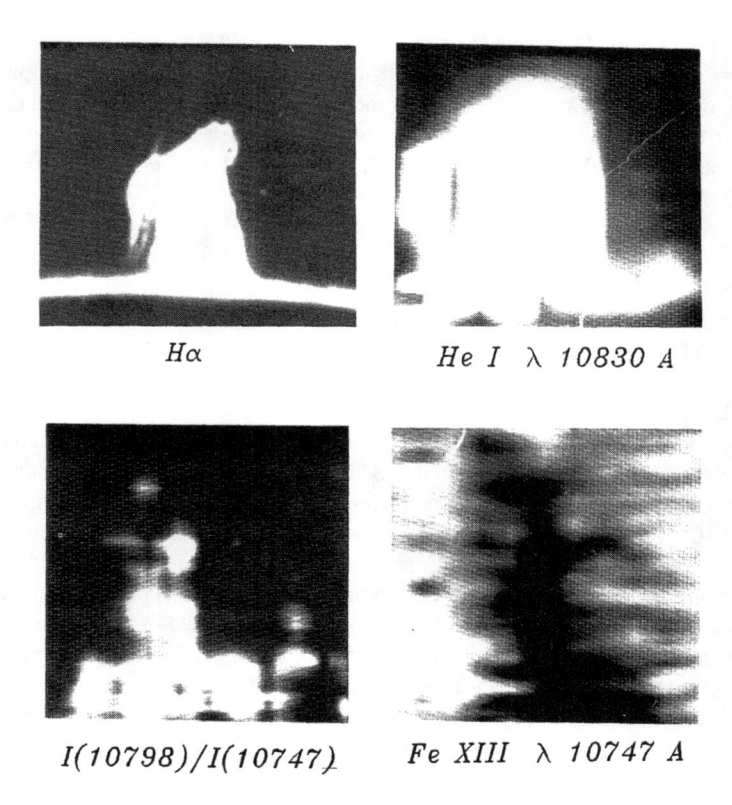

He I λ 10830 A

I(10798)/I(10747)

Fe XIII λ 10747 A

Figure 3.14. Hα and He I λ1083Å images of an active prominence region observed at the west limb of the Sun on 7 November 1986. The coronal cavity is seen as reduced emission in the coronal line Fe XIII λ10747Å. The ratio of the two Fe XIII lines, which is proportional to N_e, shows an enhancement of electron density close to the prominence (Courtesy J.C. Noëns, Pic-du-Midi)

where A_{EL} is the element abundance, ν the frequency, <g> the effective Gaunt factor and f the oscillator strength for the particular line. R_i (T) is the ionization fraction – a known function of temperature for coronal conditions. Engvold et al. (1987) studied the P-C interface using this formalism and the line intensities measured from nine slit positions in three prominences observed with the ATM/Skylab (Kjeldseth-Moe et al. 1979). The F(T) functions derived for the different positions in the prominences are all very similar. The average function F(T) (Figure 3.16) has the same shape as that of the quiet Sun, but the values of the two differ by approximately a factor 3.

CHAPTER 3: PROMINENCE ENVIRONMENT 63

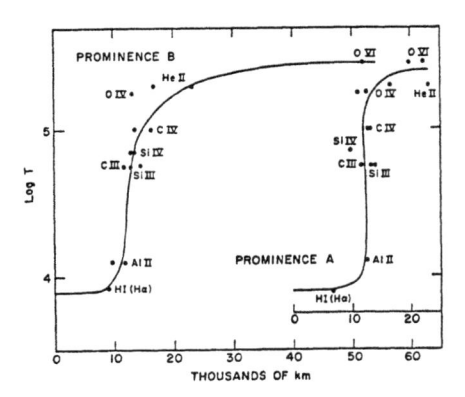

Figure 3.15. The temperature profile through the P-C transition region obtained from observation with a slitless spectrograph during the eclipse of 7 March, 1979 (Schmahl 1979).

The geometrical structure of the P-C transition region is calculated by extracting the temperature gradient from the empirical function F(T) (Eq. 3.4). The derivation of a model atmosphere requires (i) a value of the electron pressure at some reference temperature, and (ii) the fraction of the effective emitting area (A_{eff}) as a function of temperature.

The important terms of the energy balance in the P-C transition region are given in Equ. (3.6) for the case of steady state conditions.

$$E_H = E_R + \nabla \bullet F_c + \nabla \bullet F_v \qquad (3.6)$$

E_H is the rate of mechanical and ohmic heating, E_R radiative losses, F_c conductive energy flux, and F_v energy flux associated with mass flows.

When thermal conduction is the dominant term in the energy balance, we may assume a constant conductive flux F_c and write

$$F_c = 1.1 \times 10^{-6} \, T^{5/2} \, (dT/dh). \qquad (3.7)$$

We combine this equation and Equ. (3.4) to obtain

$$\log F(T) \simeq \text{const} + 5/2 \log T. \qquad (3.8)$$

It is seen from Figure 3.16 that this relation appears to be a reasonable approximation for $T \geq 100\,000$ K. Hence, thermal conduction seems to play a dominant role in the high temperature regions of the transition zone. At lower temperatures deposition of energy is needed to fit the empirically derived temperature gradient.

Schmahl (1979) and Engvold et al. (1987) find that the rise of temperature in the P-C interface is less steep than the C-C transition.

3.5.3 A FRAGMENTED AND DYNAMIC TRANSITION REGION

The temperature and pressure structure of the P-C interface is the result of the reciprocal action of mass flows, transport and deposition of energy (thermal conduction, mechanical heating, etc.), and radiative cooling in the region. The orientation of the local magnetic field plays a dominant role in shaping and structuring the transition sheath. Based on a study of UV lines Poland and Tandberg-Hanssen (1983) conclude that prominences and the P-C interface consist of numerous isothermal flux tubes of various temperatures. The alternative approach which is elaborated above is to assume that all flux tubes have the same cool core and a transition shell along the tube.

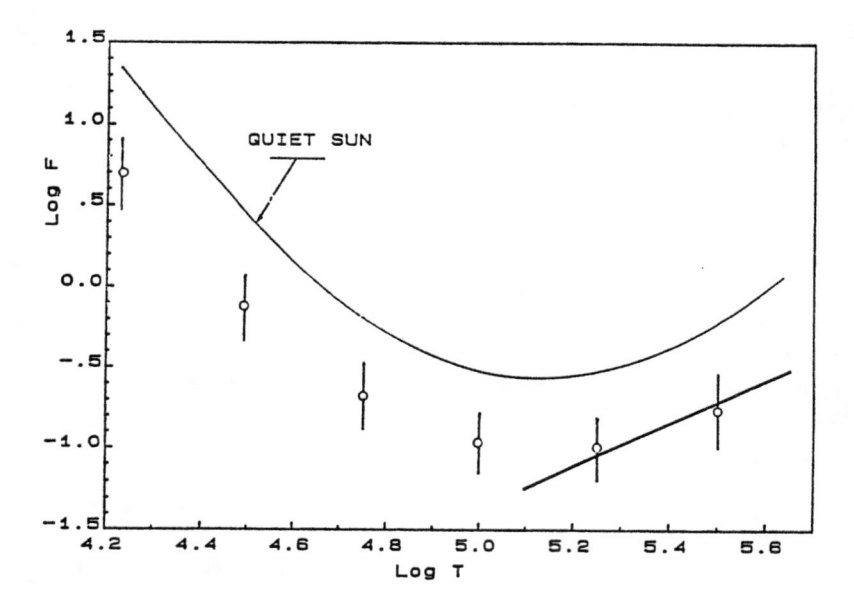

Figure 3.16. Average emission measure function F(T) of three quiescent prominences observed with ATM/Skylab. The upper function represents the quiet Sun. The straight line shows the relation predicted by thermal conduction alone; Equ. (3.8). The vertical bars give the rms variation from the nine individual observations of the P-C transition region (Engvold 1987).

The intricate and pervasive fine-scale structures of the prominence and transition region (cf. Orrall and Schmahl 1980) clearly contain important clues to the heating mechanisms and various forms of engergy transport in the transition region. The transition region plasma is evidently controlled by dynamic effects. (Pneuman Kopp (1978) discuss flows in the C-C interface.) Flows will dominate the plasma if the enthalpy flux is comparable to or exceeds the conductive flux.

Conversely, the "conductive" regime of a flowing plasma may be estimated by setting

$$v \times E < F_c \tag{3.9}$$

where v is the flow velocity and E is the thermal energy density of the gas. Let us examine the conditions in the P-C transition region at, say, $T \simeq 10^5$ K. Using Equ. (3.4) one may replace the temperature gradient in Equ (3.7) to obtain

$$F_c = 3.48 \times 10^6 \, A_{eff} \, P_e^{\,2}/F \tag{3.10}$$

Taking $E = 3P_4$, and the fact that $F \simeq 0.11$ at the temperature level in question for quiescent prominences (see Figure 3.16), and combining Equ. (3.9) and (3.10) one derives

$$v_{critical} \leq 10^7 \, A_{eff} P_e \,. \tag{3.11}$$

Since $0.05 \leq A_{eff} < 1$ and $P_e \simeq 0.01 - 0.1$ dyn cm^{-3}, one finds that the critical flow speeds in the P-C transition region would be $v \leq 5$ km s^{-1}. Line-of-sight velocities in this range and larger are, however, commonly observed in quiescent prominences (Chapter 2) and should also be expected in the transition region. One must simply conclude that plasma flows cannot be ignored in modelling of the P-C transition region.

If the transit time of the gas through the interface region is shorter than, or comparable to, the time scales for the slowest atomic processes then the plasma will not be ionization equilibrium. Flow velocities > 1 km s^{-1} across a steep temperature gradient is large enough to affect the ionization equilibrium of the plasma of the P-C transition (cf. Joselyn et al. 1979).

Diffusion of energetic electrons into the low temperature region of the chromosphere-corona interface from the hot corona may affect the ionization balance (Roussel-Dupré 1980; Shoub 1983). Such thermal diffusion may lead to an enhanced heat flux relative to the pure Maxwellian conditions in the steep transition regions.

No consistent modelling with stationary flows has yet been made for the P-C transition sheath. Further investigation of the fragmented and dynamic character of the P-C interface is restricted by observational limitations. Information about the underlying physical processes remains hidden in the unresolved fine structure (velocity amplitudes, flow velocities, and area filling of emitting matter).

3.6 Prominences and Environment

3.6.1 MAGNETIC FIELDS AND CHROMOSPHERIC STRUCTURE

Various methods used to measure prominence magnetic fields and the results from numerous studies over recent years are discussed by Leroy (1987). The magnetic field vector in quiescent prominences is found to be basically horizontal, and, on the average, enters and exits the prominence at an angle of about 25 degrees with the long axis of the prominence (see Chapter 4). It is believed that the sheared magnetic field is a fundamental topological feature in the formation and existence of prominences.

The Hα fine structure potentially provides the finest chromospheric vector magnetograph we are ever likely to have. Off-band Hα filtergrams of a large, dark filament showing the underlying disk structure give evidence for a horiziontal magnetic field which is inclined at a small angle to the filament (Zirin 1966). Foukal (1970) finds that the streaming direction of fine-scale fibrils reverses across the neutral corridor of active region filaments.

McCabe and Mickey (1981) detected an enhanced chromospheric brightness in the HeI λ10830Å line filament channels. The helium lanes also appear where there is no evidence for an overlying Hα filament. The brightness increase is actually a weakened λ10830Å absorption as a result of increased ionization of helium by absorption of EUV radiation from the helmet streamer region. The mechanism is discussed in detail by Milkey et al. (1973), and by Zirin (1975).

In their comprehensive study d'Azambuja and d'Azambuja (1948) investigated the bright narrow margin that is often seen at the base of Hα filaments, e.g. in the centerward side when the filament is located near the solar limb. The phenomen is noticed in the photograph from the Norikura Solar Osbervatory shown in Figure 3.9. The bright margins represent a brightness temperature larger than that of the quiescent chromosphere, contrary to the filaments themselves. d'Azambuja and d'Azambuja concluded that the margins are part of the prominence rather than being localized in the underlying chromosphere. Others claim that the phenomenon is located in the chromosphere at the foot of prominences (Zirin 1966). The phenomenon may represent enhanced local heating possibly as the result of reconnecting magnetic fields (see Section 3.7.2).

3.6.2. ASSOCIATION WITH SUPERGRANULATION NETWORK

Quiescent prominences often appear with loop-like lower boundaries and a filamentary or choppy main body (see Figure 1.2; Chapter 1). Furthermore, the arch footpoints of semiregular spacing seems to be associated with the supergranulation network. Plocieniak and Rompolt (1972) found that the "legs" of quiescent prominences are rooted in adjoining cell boundaries; the so-called downdrafts (Figure 3.17). These are known to be locations of flux concentrations in the photosphere. The photospheric magnetic fields are subject to dispersion by the action of the supergranulation flows. The average rate is 140 m s^{-1}, corresponding to about 330 000 km per rotation of the Sun. If the anchor points of the supporting magnetic fields of prominences move about at such speeds also the filament channels would be subject to motion and dispersion. The typical lifetimes of ≥ 2 hours for large-scale prominence legs may be an effect of local rearrangements of the magnetic fields taking place through the action of the supergranulation flows. A cursory inspection of the synoptic charts of the Meudon Observatory shows that the shapes and relative positions of the filaments (channels) are stationary for several solar rotations contrary to the assumed dispersive action of the supergranulation flows. Long-lived filaments are in fact used as tracers for measurements of solar rotation. Soru-Escaut et al. (1985) noted that particular points along long-lived quiescent filaments exhibited a rotational rate which was the same as the Carrington rate (cf. Schröter 1985). These locations are independent on latitude and are referred to as "pivot points". Mouradian et al. (1987) found that new magnetic flux often emerges close to pivot points. This cirumstance and the apparently rigid rotation of pivot points suggest that the

CHAPTER 3: PROMINENCE ENVIRONMENT 67

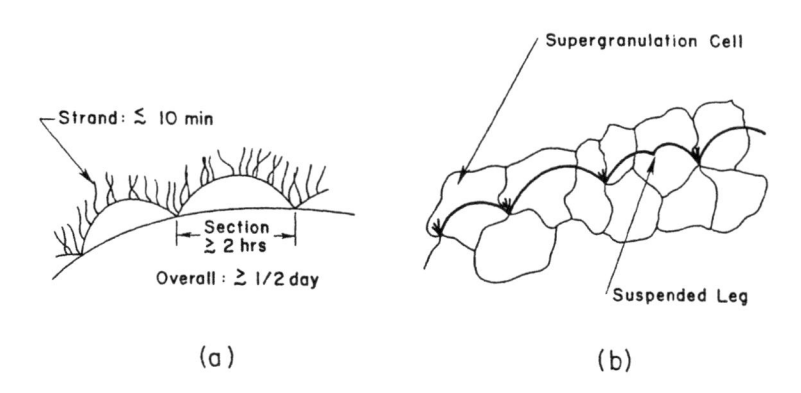

Figure 3.17. (a) lifetimes of prominence features and the assumed location of prominence legs with respect to supergranulation cell boundaries (T. Forbes, 1986).

phenomenon is rooted deep inside the Sun. It is presently unknown whether pivot points are associated with particular prominence structures.

The filament channels maintain their identity and positions in spite of the reshuffling of the magnetic fields that continuously takes places in the photosphere. This suggests that filament channels are the result of a subtle, but stable, large-scale action. (The systematic poleward migration of high latitude filament channels was mentioned in Section 3.4.2). An agent that could give rise to the observed large-scale magnetic topology of filament channels could possibly be slow large-scale flows in the photospheric and sub-photospheric layers. It has been speculated whether giant cell motions would give rise to systematic movements of the bases of the suppo ring magnetic arches in opposite directions on the two sides of the neutral line (Schmieder et al. 1984). The idea was put forward to explain the apparent vertical motion of many quiescent prominences (see Malherbe and Priest 1987).

3.6.3 DYNAMICS

Observations of the helmet streamers and prominence complex bear out both large-scale motions and micro-motions. The remarkable association of dark filaments with velocity reversal lines was first noticed in Dopplergrams made with the UVSP experiment on the SMM satellite (Schmieder et al. 1984b; Engvold et al. 1985; Athay et al. 1986; Malherbe et al. 1987). The Figures 3.18 and 3.19 show large solar filaments and their associated line-of-light flows observed in the C IV $\lambda1548$Å line. One also finds a significant component of the flow parallel to the filament.

The "cool" filament matter exhibits noticeable mass motions in the vertical direction (Mein 1977; Martres et al. 1981; Malherbe et al. 1983; Kubota and Vesugi 1986; Engvold and Keil 1986). The dominant direction of the flow is upwards at speeds of 1–3 km s^{-1}. It is not known whether the observed motion represents a true vertical flow or if it is rather an effect of radiative transfer in a "convective" or oscillating plasma (Engvold and Keil 1986). For more details on mass flows the reader is referred to Section 2.3.

3.6.4 THE MASS OF CORONAL CAVITY AND PROMINENCE

Saito and Tandberg-Hanssen (1973) found that the formation of quiescent prominences requires much more material than is available in the cavity before depletion. Using the projected area of prominences measured from the eclipse photographs the authors may have overestimated the prominence mass by a factor 10 or more. In the following we adopt the cavity dimensions from the X-ray study by Serio et al. (1978), which gave height > 50 000 km and width 60 000 km, and

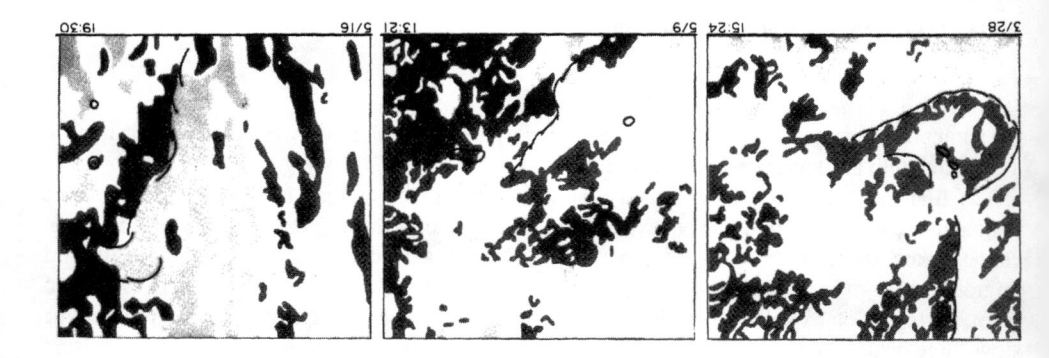

Figure 3.18. C IV λ1548A Dopplergrams obtained with the UVSP instrument on SMM. The grey shading indicates velocities in excess of 4 km/s; dark grey is blue shifted and light grey is red shifted. The locations of Hα filaments are indicated by tufted lines (Athay 1985).

a number density $N = 4 \times 10^8$ cm^{-3}. This density corresponds to a cavity pressure as low as $P = 0.05$ dyn cm^{-2} at $T = 10^{-6}$ K. A compression of the cavity plasma by a factor 100, which would equal the generally accepted values for prominence density according to Hirayama (1985), results in a volume of a slab prominence of height 30 000 km and thickness 500 km. Since only part of the prominence volume is filled with high density matter (note the prominence fine structure), we find that the resulting required slab thickness becomes ≥ 500 km. It may therefore be concluded that the matter contained in the cavity region is sufficient to account for the mass of a prominence of height 30 000 km and of thickness comparable to the observed size of the prominence small-scale structure.

CHAPTER 3: PROMINENCE ENVIRONMENT 69

SMM/UVSP 10 AUG 1980

C IV Intensity

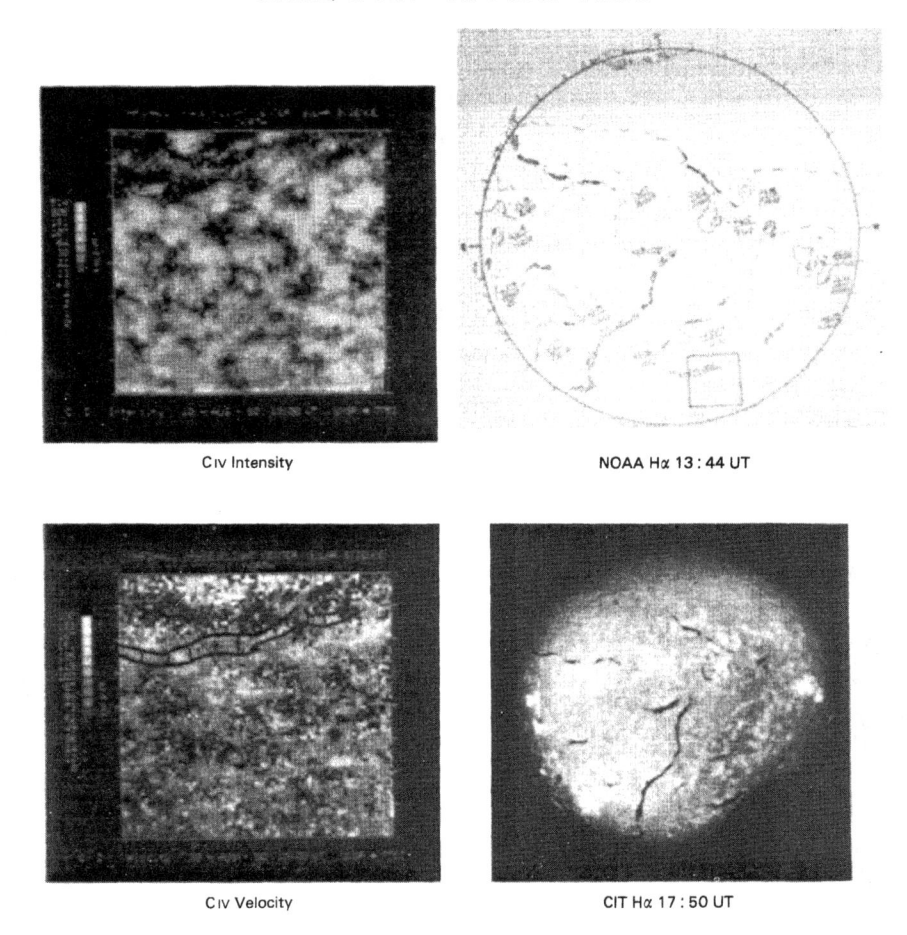

NOAA Hα 13 : 44 UT

C IV Velocity

CIT Hα 17 : 50 UT

Figure 3.19. Left: C IV intensity and velocity maps obtained from observation with the UVSP flown onboard SMM. Right: The positions of Hα filaments are seen from the NOAA neutral line drawing and the Hα filtergram (Engvold et al. 1985).

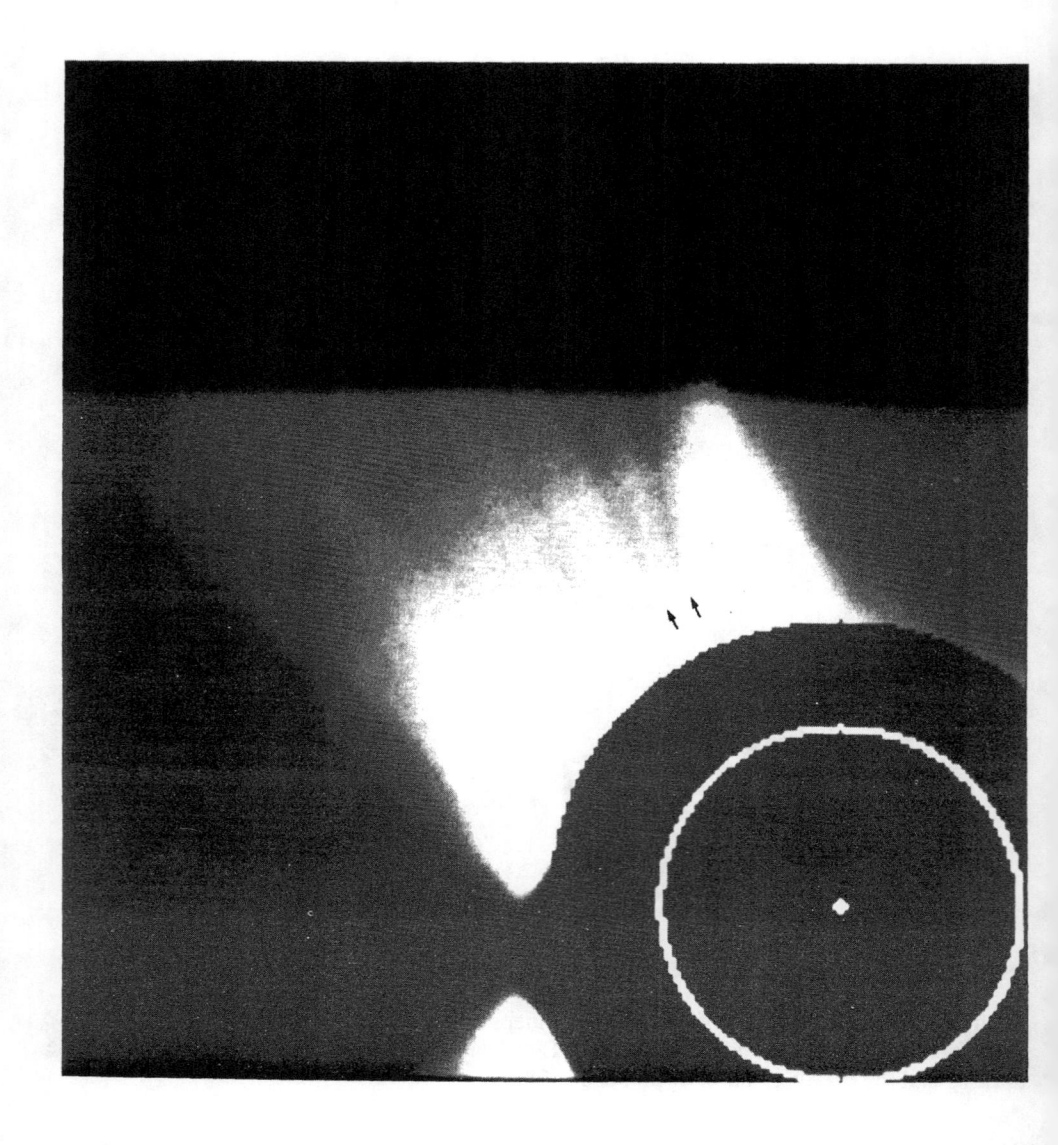

Figure 3.20. Coronagraph images showing two coronal voids (arrow heads). The image is treated digitally to enhance the contrast (Courtesy R.M. MacQueen).

3.6.5 CORONAL VOIDS - A SOURCE OF PROMINENCE MASS?

While examining white-light coronagraph images MacQueen et al. (1983) noticed the presence of thin, dark, ray-like structures extending through the corona as far as $3R_o$ from the limb. The width of the structure changes only slowly with radial distance. The structures are referred to as coronal "voids" due to their low brightness in white-light coronal images (Figure 3.20).

The low-level white-light brightness from the void is evidently caused by reduced Thomson scattering which then suggests a particularly low electron density in the region. It is assumed that a coronal void represents a temporary reduced electron density in the midst of the more usual coronal conditions.

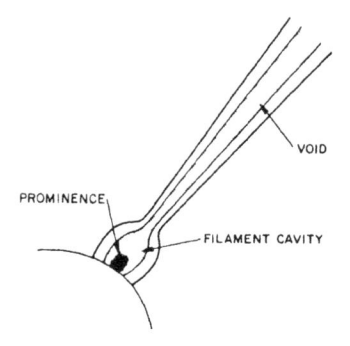

Figure 3.21. A speculative geometry indicating the possible extension of a filament cavity into a coronal void in the absence of closed magnetic field above a prominence (MacQueen et al. 1983).

MacQueen et al. (1983) found that voids are spatially correlated with magnetic neutral lines, and, more significantly, with prominences (Figure 3.21). The physical nature of voids is therefore of considerable interest and relevance to theories for prominences.

It was pointed out by MacQueen et al. (1983) that voids appear and disappear over less than 4 hours. Their lifetime is short compared to the lifetime of filament channels and prominences. It is conceivable that a void may reflect conditions which are appropriate for some particular phase in the evolution of prominences. MacQueen et al. noticed that the appearance of 31% of the voids in their material could be tied to the birth of a prominence. This observation cannot be attributed to chance coincidence. The result suggests that material which initially is situated within the void boundary may provide the necessary mass for the formation of a prominence. Provided the magnetic energy density of the structure dominates the ambient corona, a sudden reduction in the base temperature (and pressure) could trigger an instability and subsequent collapse that may empty the void region.

3.7 Modelling of the Helmet Streamer/Prominence Complex

3.7.1 HELMET STREAMER AND CAVITY

Pneuman (1968) analyzed the helmet streamer magnetic configuration consisting of a central weak field (helmet) and a stronger open, outer field (streamer). The energy flux entering the weak field region at the base is dissipated and subsequently radiated away to permit a stable equilibrium condition. In the outer stronger field region the incident waveflux is not dissipated, and the result is an open structure (Figure 3.22).

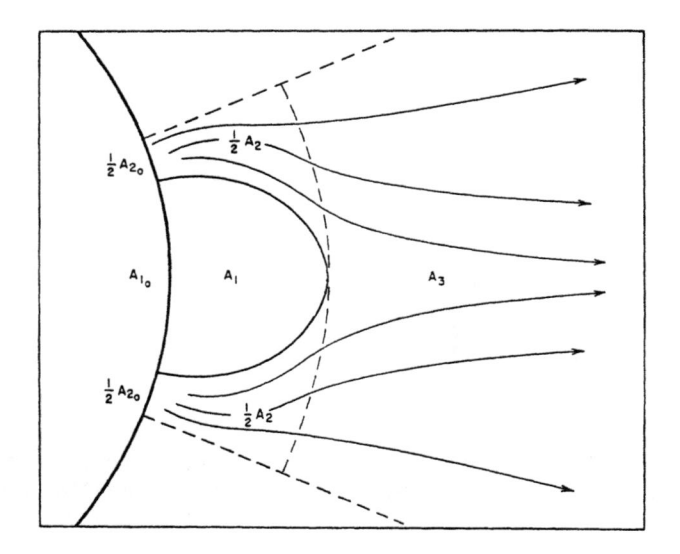

Figure 3.22. Geometrical sketch of coronal streamer configuration. A_1 is the cross section of the helmet, A_2 the cross section of the open region adjacent to the helmet (Pneuman 1968).

An et al. (1985) have analyzed the formation of a coronal cavity in the magnetic configuration of the helmet streamer structure using time dependent, ideal MHD equations (Figure 3.23). They concluded that the open field regions of the streamer structure is a source of solar wind. Such regions of upflow are actually seen around large quiescent filaments (Engvold et al. 1985). In the closed field regions the plasma may attain a steady, and possibly also stationary, state, according to An et al. Deeper inside the helmet the hot coronal plasma will condense and result in a drainage, which subsequently form a cavity. The condensation may lead to prominence formation if the local magnetic field topology provides the necessary support. Otherwise, there will be a cavity.

CHAPTER 3: PROMINENCE ENVIRONMENT 73

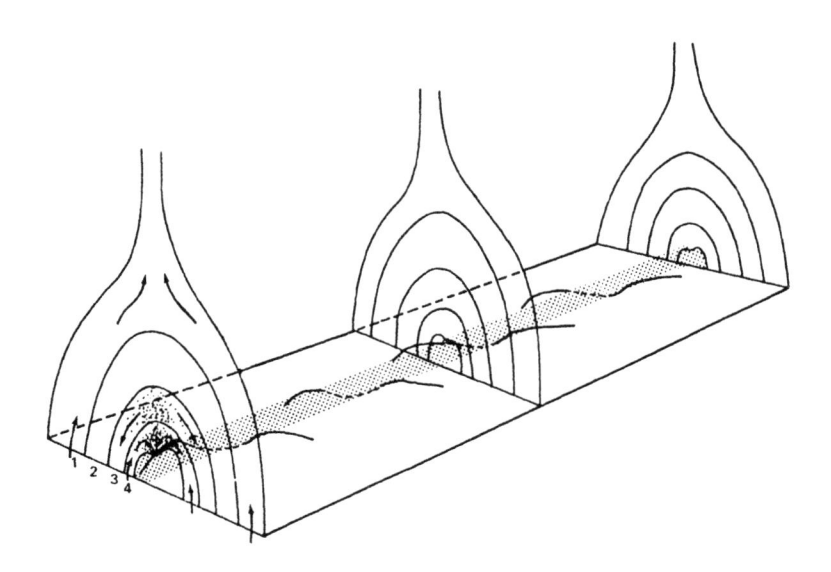

Figure 3.23. A 3-D description of a coronal streamer and its inner structures. The magnetic fields for the coronal cavity and the quiescent prominence are depicted as a global dipole and a sheared local field, respectively. The direction of anticipated plasma flows are represented by arrows (An et al. 1985).

3.7.2 MAGNETIC FIELD TOPOLOGY

It was pointed out by Tang (1987) that quiescent prominences favour neutral lines between bipolar regions over neutral lines of bipolar regions. To be in accordance with this fact prominence modelling should prescribe magnetic field topology evolving from reconnection of field lines of two adjacent bipolar regions.

The observed helmet streamer configuration cannot yet be used to distinguish between magnetic field configurations with normal or inverse magnetic polarity. Various developments of the Kippenhahn-Schlüter (1957) model (see Wu and Xiao 1986; Ballester and Priest 1987) would possibly fit into the helmet structure. The general picture of quiescent prominence being trapped by loop-like magnetic structures of the cavity regions is not suggestive of field configurations with a neutral sheet as prescribed by the Kuperus-Raadu (1974) model. The transient coronal void phenomenon could, however, provide interesting possibilities for Kuperus-Raadu models.

Pneuman (1983) has argued that bipolar magnetic regions such as those present in coronal cavities may develop into closed configurations that can actually support prominences. A "figure 8" configuration of the field may develop as the result of slow motions of the foot-points in the bipolar regions or in an emerging and rising field. The lateral Lorentz force will not necessarily be balanced by gas pressure at all heights. One might attain plasma $\beta > 1$, which could lead to an inward collapse of the field structure and magnetic reconnection (Figure 3.24). The closed loops will actually be helices, because of the initial shear in the magnetic fields, within which the cavity matter will cool radiatively and form a prominence. The question of mass is not necessarily a problem in Pneuman's model since one may argue that the cavity regions possibly contain enough matter for prominence formation (cf. Section 3.6.4). The observed bright narrow margins at the base of Hα filaments could be the observable results of magnetic reconnection at the bottom parts of the helices.

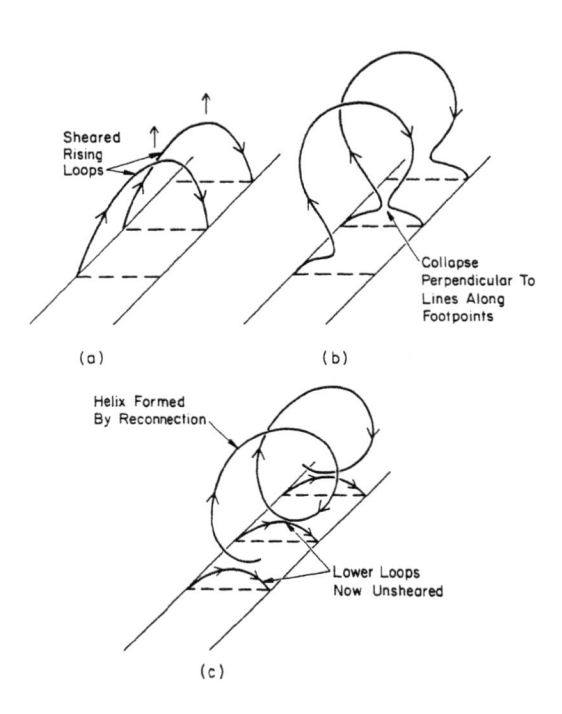

Figure 3.24. Schematic illustration of a rising, sheared bipolar region which may collapse inward to produce isolated helices above the neutral line. The prominence is formed in the bottom central part of the helices by condensation of cavity material and subsequent downflow (Pneuman 1983).

3.7.3 SIPHON-TYPE MODELS

Several siphon-type models have been analyzed in order to explain the assumed transport of mass into prominences (Pikel'ner 1971; Engvold and Jensen 1977; Priest and Smith 1979; Ribes and Unno 1980; Poland and Mariska 1986). The entire siphoning magnetic tubes must necessarily be contained in the cavity region inside the helmet structure. The objective of siphon-type models is to explain the presumed transport of matter into the prominence (Figure 3.25). There is presently no clear observational evidence for a mass flow into the prominence region and, furthermore, the

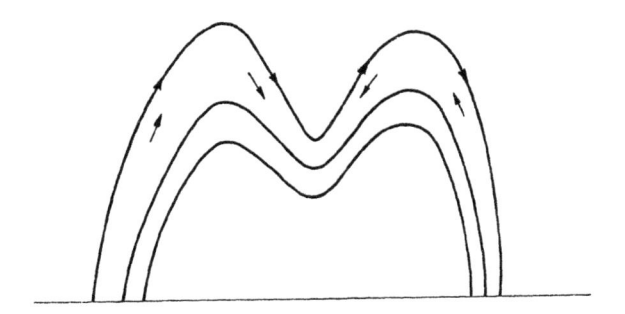

Figure 3.25. Magnetic configuration for a stable syphon type prominence suggested by Pikel'ner. The "cool", condensed material is accumulated in the pit in the arch (Pikel'ner 1971).

actual need to supply prominence mass from the adjoining corona, or from the chromosphere below, is questionable and subject for discussions and further investigations.

An et al. (1987) studied the development of a pit in a magnetic structure with normal polarity assuming symmetric injection of matter into initially potential dipolar magnetic loops imbedded in a hydrostatic atmosphere. The injection of matter was suggested to be the action of spicules trapped in closed magnetic loop systems, or the effect of some as yet unspecified flows, but the pit only forms for unreasonably large values of the coronal plasma β.

3.8 Conclusions

Studies of global and micro-scale structures of prominences and their environment define boundary conditions for modelling of prominences and neighbouring regions. It must be underlined that major questions still remain essentially unanswered. It appears that the mass contained in the cavity region alone is possibly sufficient to form prominences. This fact does not, however, rule out that prominence mass is being supplied from the corona (during the presence of coronal voids?) or from the chromosphere below. It is not known if the apparently systematic flows in the 100 000 K plasma surrounding prominences are connected to the vertical mass motions observed in prominences. One should note that the vertical motions in quiescent prominences may not represent an actual flow of matter since the magnetic field is basically horizontal.

The following questions concerning the environment are fundamental to a better understanding of prominences:

- What is the significance of the sheared magnetic fields and velocities associated with prominences and cavities?
- Is the local field topology of the cavities, i.e. the amount of shear, the prime parameter that determines whether or not a prominence is formed?
- Does the magnetic field topology exist prior to the condensation of the cool matter, or is local accumulation of matter shaping the magnetic field into a supporting structure?
- Is there a photospheric flow pattern associated with filament channels which produces the necessary magnetic field topology?
- How much of the matter of the cavity region exists at low temperatures?
- Is the large-scale flow of the P-C interface and cavity region responsible for the formation and maintenance of quiescent prominences?
- Are prominence thermally and dynamically shielded from the surrounding hot corona by the magnetic arches of the helmet streamers?
- Is there an association between the magnetic flux concentration in the photosphere and the magnetic fields of prominences?

Acknowledgements.

The author acknowledges helpful discussions and contributions to this review by E. Hiei, E. Jensen, S.L. Koutchmy, J.-L. Leroy, R.M. MacQueen, H. Morishita, J.C. Noëns, E.R. Priest, T. Tsubaki, and H. Zirin.

CHAPTER 4

OBSERVATION OF PROMINENCE MAGNETIC FIELDS

J.L. LEROY
Observatoire du Pic-du-Midi et de Toulouse
14, Avenue Edouard Belin
31400 Toulouse - France

4.1 Historical Steps

In his comprehensive book on solar prominences Tandberg-Hanssen (1974) wrote : "The single, physically most important parameter to study in prominences may be the magnetic field. Shapes, motions, and in fact the very existence of prominences depend on the nature of the magnetic field threading the prominence plasma..."

Twelve years later this statement is still pertinent and for any solar astronomer the central role of the magnetic field in prominence phenomena is obvious. Therefore, it is very important to perform good quality magnetic field measurements both in the prominence plasma and also in the neighbouring solar atmosphere (which is possible in the photosphere but much more difficult in the corona!) The story of magnetic field measurements related to prominences begins with the early discovery, quickly following the achievement of the Babcock magnetograph, that quiescent prominences are located above the neutral lines of the longitudinal photospheric field (Babcock and Babcock, 1955). This very significant result will be discussed later in more detail.

Aiming a magnetograph towards the prominences which pass by at the solar limb was quite natural although not so easy to achieve practically. Nevertheless, Zirin and Severny succeeded in providing in 1961 the first magnetic field determinations : in quiescent prominences the field strength was found to be of the order of 50 Gauss (0.005 Tesla) a figure which is slightly larger than most values provided by modern measurements. That was the beginning of actual magnetic measurements (rather than estimates based on motions or other properties) and reliable data were to become available soon.

The next stage was the design of a coronagraph specially suited for the study of prominences. This task was tackled with great skill at High Altitude Observatory (H.A.O.) and resulted in a fine coronagraph-magnetograph (Lee et al., 1965) which enabled the observation of solar prominences above the solar limb with a reduced amount of scattered light. An ingenious device allowed an accurate calibration of Zeeman displacements, even though prominence lines have very different shapes and widths. Most observations were performed with the help of the Hα line. This instrument was used extensively and provided the wealth of data which can be found in the theses of Rust (1966) and Harvey (1969) and related papers, as well as in further publications by Tandberg-Hanssen, Malville and others (see Tandberg-Hanssen, 1974).

In the following years other attempts to measure prominence magnetic fields were made especially in the Soviet Union. Various instrumental schemes including sometimes very simple devices (photography of two spectra behind a Wollaston prism) were proposed, with rather variable results. In contrast with the results of the H.A.O. instrument, some investigators derived a rather strong prominence field, even in polar prominences (Smolkov, 1971). For some time this troublesome contradiction (Wiehr, 1972) cast some doubt on the true value of the field strength in quiescent prominences (Rust, 1972). Later on, it became clear (Bashkirtsev, 1975; Bashkirtsev and Mashnisch, 1976) that instrumental polarization was responsible for the fictitious high field values reported by this Soviet group; such difficulties stressed the advantage of on-axis instruments with no instrumental polarization for magnetographic investigations.

In spite of the sophistication of magnetographs, only the longitudinal Zeeman effect could be measured by the preceding instruments and while a good diagnostics of the field strength had been obtained (Tandberg-Hanssen, 1970) only preliminary results concerning the field orientation were available (Tandberg-Hanssen and Anzer, 1970). The next progress was to come from another, rather obscure, spectroscopic effect, namely the Hanle effect.

It is well known that resonance scattering results in partially polarized radiation, which has interesting astronomical consequences (Ohman, 1929). The modifications of resonance scattering which occur in the presence of a magnetic field had been discovered in the laboratory (Hanle, 1924) at the time when Lyot (1934) measured anomalous (non-tangential) polarizations in the monochromatic emissions of prominences; nevertheless, the link between both discoveries was not established and the tentative explanation of Lyot's observations was not adequate (Thiessen, 1951). During the next decade the diagnostic content of the Hanle effect was recognized by Hyder (1965a) but a correct theory of the Hanle effect became available only some years later (House, 1971). In the time being, new observations of the polarization of prominence emission lines had been performed (see a review in Leroy, 1979) and it was obvious that substantial work in this direction would provide interesting results on the magnetic properties of prominences.

This program was engaged by two teams: a collaboration between H.A.O. and Sacramento Peak Observatory resulted in the construction of two successive "Stokesmeters" (Baur et al., 1980; Baur et al., 1981) and the theoretical side of the interpretation was also advanced (Landi degl'Innocenti, 1982). On the other side, a polarimeter was built at Pic du Midi Observatory (Ratier, 1975) and a systematic study of Hanle effect theory was undertaken at Meudon Observatory (Sahal-Bréchot et al., 1977 and further papers; see Bommier, 1987). Frequent contacts between both groups allowed intercomparisons, increasing the confidence in this new method of magnetic field measurements.

While two dozen papers resulted from the investigations based on the Hanle effect, two other teams in the Soviet Union tackled again the Zeeman measurement of prominence magnetic fields. Under the direction of the late Prof. Nikolsky a spectrally-scanned magnetograph was built and set at the focus of the large coronagraph of Kislovodsk (Nikolsky et al., 1982). Although this method is able to yield only one component of the field, it is certainly valuable to compare the results provided by the Hanle method and by the Zeeman method. Useful Zeeman measurements were also provided by another group working with the magnetograph of Sibizmiran (Bashkirtsev and Mashnich, 1980) after the initial instrumental difficulties had been overcome.

Thus, one can see that a major effort involving astronomers from several countries has been invested during the last ten years. The volume of observations is considerable (at Pic du Midi alone, 400 prominences have been studied) and it is worth now giving an assessment of the output of these programs aimed at a better knowledge of prominence magnetic fields. In the following we concentrate on the recent results, taking as a starting point the Oslo Colloquium on solar prominences where former work was reported.

4.2 Investigations based on the polarimetry of spectral lines

4.2.1 THE ZEEMAN EFFECT

It is not useful to give here one more report on the Zeeman effect : this topic is treated in many comprehensive papers (see for instance the recent reviews by Stenflo (1985a and b) and, in the particular case of solar prominences, the description by Tandberg-Hanssen (1974)). Thus, we point out only some specific questions linked to prominence studies.

When emitting atoms are submitted to a magnetic field directed along the line of sight we observe the so-called longitudinal Zeeman effect : the spectral line is split into two components (in the simplest case of "normal" Zeeman splitting) with opposite circular polarizations. However, for a field strength of less than 100 Gauss (0.01 Tesla) which is characteristic of prominences, the splitting is far from complete because the Doppler width of spectral lines is larger than the Zeeman splitting $\Delta\lambda z$ ($\Delta\lambda z \simeq 10^{-4}$ nm, in the visible spectrum, for a field of 100 Gauss). In such a case, which is the weak field approximation, one can show that the Stokes parameter V

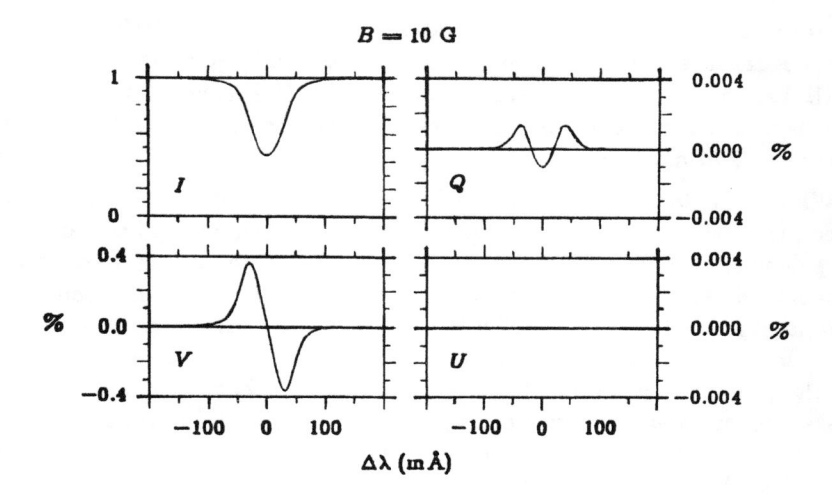

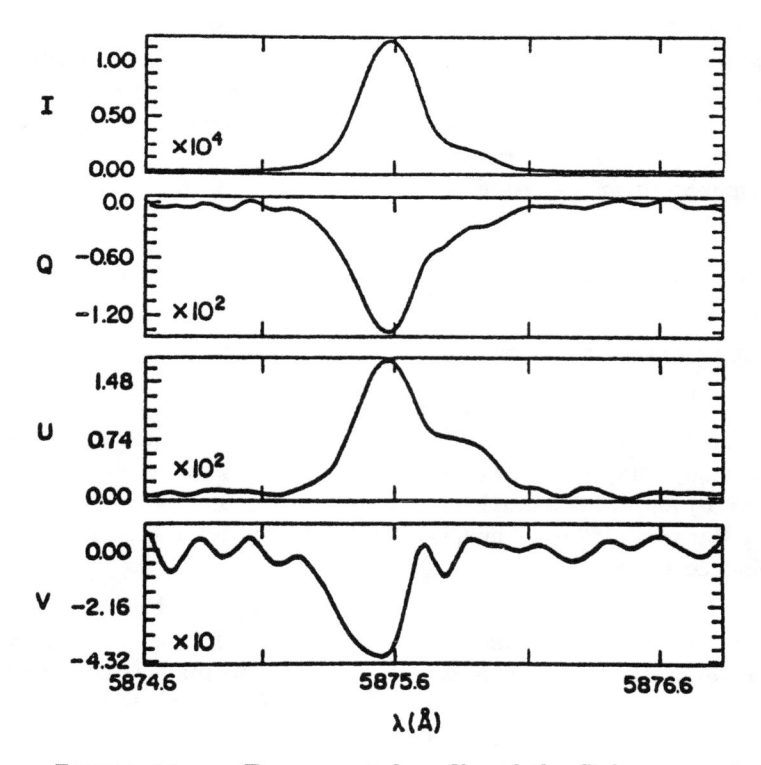

FIGURE 4.1. — Top: computed profiles of the Stokes parameters resulting from the Zeeman effect in the λ 5250 photospheric line (Stenflo, 1985a). Bottom: observed profiles of Stokes parameters resulting from the Hanle effect in the λ 5876 prominence line (Querfeld et al., 1985).

(describing the circular polarization of light) has a typical S-shaped variation across the line profile according to the expression :

$$V = K_{//} \, B_{//} \, \frac{dI}{d\lambda} \tag{4.1}$$

$K_{//}$ is a constant and $I(\lambda)$ describes the intensity variation across the line profile. This formula is valid even though the field vector is not purely longitudinal, V being proportional to the projection $B_{//}$ of the field vector on the line of sight.

The transverse Zeeman effect (field vector perpendicular to the line of sight) results in the appearance of a Zeeman triplet with linearly polarized components. In the customary case of incomplete splitting, the Stokes Q and U parameters, which describe the linear polarization of light, are :

$$Q = K_\perp \, B_\perp^2 \, sin2\chi \, \frac{d^2 I}{d\lambda^2} \tag{4.2}$$

$$U = K_\perp \, B_\perp^2 \, cos2\chi \, \frac{d^2 I}{d\lambda^2} \tag{4.3}$$

$K_\perp$ is a constant and the angle χ depends on the angular reference system of the polarimeter. Typical shapes of Q, U and V profiles across a normal Zeeman triplet are displayed on Figure 4.1: one can notice the strong λ - dependence whence the need for a good spectral resolution in Zeeman measurements.

If we could use directly the expressions (4.1), (4.2) and (4.3) the magnetic field diagnostics would be only a matter of polarimetry (the determination of Q, U and V). However, the actual situation is more intricate and, first of all, (4.1), (4.2) and (4.3) are valid only when the field under study is homogeneous. If the spatial resolution of the magnetograph is not adequate new problems arise: during the last decade, several thorough investigations of the photospheric magnetic field have revealed that the interpretation of measurements performed with insufficient spatial resolution is extremely difficult, especially if we try to perform complete field vector diagnostics (Stenflo, 1985b). Remember also that when prominences are observed over the solar limb the line of sight integration results in the mixing of different emissions in one single measurement, even though the angular resolution is excellent.

As a matter of fact, complete vector field diagnostics are generally beyond instrumental capabilities: given the functional form of the expressions (4.1), (4.2) and (4.3) it is clear that the square power of the field strength $(B_\perp^2)$ yields very small quantities for Q and U as soon as B is small (see Figure 4.1, top). It follows that, for a given instrument, the threshold of measurements is much less favourable for the transverse field component than for the longitudinal component: representative figures are 1 Gauss for $B_{//}$ and 50 Gauss for $B_\perp$ which means that in solar prominences the Zeeman diagnostics can provide (and has actually provided) $B_{//}$ only.

A serious problem with prominence magnetic field measurements results from the limited choice of spectral lines available for a good polarimetric analysis: only Hydrogen, Helium and a few metallic lines are bright enough to allow useful measurements. It means that we cannot choose, as in photospheric studies, specific lines with a particular splitting (for instance, a very large splitting). Rather, for the lines which are available, one must rewrite the expression (4.1) as:

$$V = K_{//} \, g_{eff} \, \frac{dI}{d\lambda} \, B_{//} \qquad (4.4)$$

where the effective Lande factor g_{eff} takes into account, in a more or less accurate fashion, the unresolved fine structure of the complex line under study. However, this is probably a negligible source of inaccuracy especially in view of another difficulty which is the precise determination of the quantity $dI/d\lambda$: it was the great novelty of the H.A.O. magnetograph (Lee et al., 1965) to measure $dI/d\lambda$ at the same time as the Zeeman splitting, which is much needed since spectral lines have variable shapes and widths even in quiescent prominences (Engvold and Malville, 1977). A still better, but more expensive, solution is to record the Stokes parameter V across the whole line profile in one exposure, thanks to a multichannel detector (Baur et al., 1981).

Last but not least, one has to remember that the excitation of prominence emission lines is a non-LTE phenomenon, deeply influenced by the photospheric light flux. The exciting beam is not isotropic which entails very interesting consequences (resonance scattering and the Hanle effect) which are treated in the next section. But there are also some consequences for the interpretation of Zeeman measurements which have been discussed in several papers (e.g. Lamb, 1970). A comprehensive analysis of this complex situation has been described by Landi degl'Innocenti (1982) and we reproduce his main conclusions: as a result of the direct excitation of prominence atoms the Stokes V parameter (circular polarization) of a prominence line reflects the superposition of two different variations; the first one is the S-shaped profile described by equation (4.4) which results from the elementary Zeeman effect while the second one has the same spectral variation as the line intensity profile. Therefore, the interpretation of the Stokes V profile is not so straightforward as it would seem, at least for small field strengths (B $\simeq$ 10 Gauss); for larger fields the Zeeman signature becomes dominant. Eventually, observers must be aware of the fact that passing from a V profile to a longitudinal magnetic field measurement is not so easy in the case of the weak field strengths which are found in quiescent prominences (of course, the task is still more difficult with double-slit magnetographs, when V is measured only at two points of the line profile). The selection of only those measurements which display clearly an S-shaped variation for the V parameter (Nikolsky et al., 1984) is certainly sound but it may involve a bias due to the elimination of very small field measurements.

4.2.2 THE HANLE EFFECT

The Hanle effect is the modification by a local magnetic field of the polarization due to coherent scattering in spectral lines (Sahal-Bréchot, 1981). The possibilities offered by this phenomenon concerning magnetic field diagnostics have been described in several reviews (e.g. Sahal-Bréchot, 1981; Leroy, 1985) and we recall here some essential features only. Resonance scattering in the solar atmosphere (Figure 4.2.a) yields partially polarized emissions with a polarization parallel to the solar limb (for permitted transitions). Close to the limb the polarization degree is small because the exciting beam is not far from isotropic: for instance, at 1 arc min. above the photosphere the polarization percentage p_{max} is below 5%, depending of course on the nature of the atomic levels involved in the emission process. If the emission occurs in a magnetic field one observes generally a decrease of the polarization degree (p < p_{max}) and a rotation of the polarization direction by an angle φ (Figure 4.2.b). In the classical theory the effect is assigned to the Larmor precession of the oscillator but a complete explanation of the phenomenon requires a quantum mechanical treatment which has been available only in recent years (Bommier 1977; Bommier, 1987).

The observed effect depends upon the orientation of the magnetic vector $\vec{B}$ relative to the Sun's vertical and to the observer direction. Throughout this chapter we adopt the reference frame of Figure 4.2.c: the field vector is defined by its modulus B, by the angle ψ with the vertical axis and by the angle θ between the projection of $\vec{B}$ on the horizontal plane and the line of sight. A vertical magnetic field ($\psi = 0$) has the same symmetry as the photospheric beam and cannot induce any change in resonance polarization. But, for increasing values of ψ one observes larger and larger depolarizations (p/p $_{max}$ decreases) and also a rotation of the polarization direction ($\varphi \neq 0$) which is more conspicuous if the field vector is near the line of sight ($\theta \simeq 0$ or $180°$). This behaviour can be summarized conveniently on a polarization diagram like that displayed at the bottom of Figure 4.2.

It is interesting to note that the polarization which is observed has the same spectral dependence as the line intensity profile so that a global measurement of the linear polarization is significant. Figure 4.1 (bottom) after Querfeld et al. (1985) yields actual Stokes profiles of the He I λ 5876 Å multiplet (the so-called D3 line) which has been often selected for magnetic measurements in prominences; this line has two main overlapping components with intensities in the ratio 8/1 which are easily seen on Figure 4.1. It is of interest to notice that the circular polarization (V parameter) is one order of magnitude smaller than the linear polarization (Q and U parameters): the Hanle effect concerns essentially linearly polarized light (remember that resonance scattering yields linearly polarized light! For the origin of the small amount of circular polarization in the Hanle effect see Landi degl'Innocenti, 1982).

Therefore, one spectral line yields two observable quantities, the decrease of the polarization degree p/p_{max} and the rotation of the polarization direction φ. This is

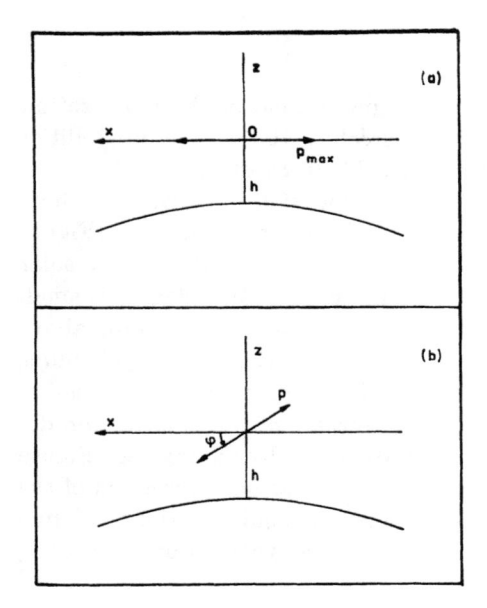

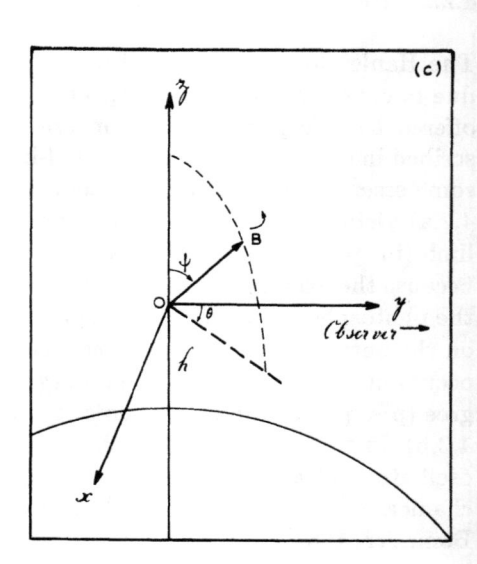

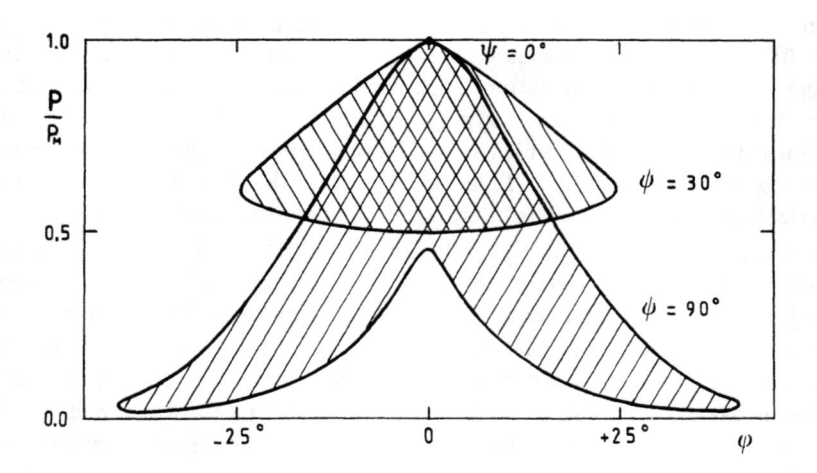

FIGURE 4.2. — Top: The polarization due to resonance scattering (a) is modified by a magnetic field (b); the reference frame pertinent to an analysis of the Hanle effect is given in (c). Bottom: Polarization diagram showing the effect on p and φ of the Hanle effect for three different values of the angle ψ.

not sufficient to determine the three parameters (B, ψ, θ) which enable a recovery of the field vector $\vec{B}$. But, observing two different spectral lines yields four observable quantities which is sufficient to achieve the diagnostics, as has been shown by Bommier et al. (1981) and by Landi degl'Innocenti (1982). As a matter of fact, the Sacramento Peak Stokesmeter has made measurements on the two components of the He I D3 line (Figure 4.1, bottom) while the Pic du Midi polarimeter measured D3 and Hβ. It must be understood that even though, from the mathematical point of view, the four observed parameters are sufficient, and even redundant, to determine $\vec{B}$, the actual situation is different because the angles ψ and θ (Figure 4.2.c) have a very different role. The magnetic depolarization is maximum for ψ around 60° and has similar (not equal) values for 90° and say 30°. As a result the observed parameters are not very sensitive to the exact value of ψ as soon as the field vector is closer to the horizontal than to the vertical ($\psi > 45°$). Eventually, given the limited accuracy of actual measurements, a precise determination of ψ is rather difficult but, in turn, a large error in ψ does not prevent a derivation of good values for B and θ. This feature, together with the property of the magnetic field of being almost horizontal in quiescent prominences, results in the fact that polarization measurements obtained with one spectral line only have been extremely useful for the investigations of quiescent prominences.

4.2.3 THE 180° AMBIGUITY

An important limitation of both the transverse Zeeman effect and the Hanle effect is the inability to distinguish two magnetic vectors which are symmetrical about the line of sight (Figure 4.3, top). The nature of this degeneracy is fundamental (Sahal-Bréchot, 1981) and linked to the character of the atom-photon interaction. The consequences have always been neglected in photospheric studies where it was implicitly assumed that the field lines ought to run directly from a positive polarity towards a negative polarity, thus allowing the selection of one out of the two solutions derived from the polarimetric analysis. This choice of normal polarity has progressively appeared to be questionable in the case of prominences and the actual geometry of prominence observations made the dilemma quite troublesome: remember that most quiescent prominences are stretched in longitude by the Sun's differential rotation so that, in typical cases, they are at an angle of about 30° with respect to the line of sight (Figure 4.3, bottom). All the early measurements showed that the vector field was consistent with a normal polarity solution having the same topology as the potential solution B_p (Rust, 1972) but it was not realized that the solution B_N with inverse polarity was equally possible. Now, if one also agrees, following Zeeman measurements (Tandberg-Hanssen and Anzer, 1970), that the angle between the prominence long axis and the field vector is small, it should become clear that the B_N solution is more likely. However, the solution of this

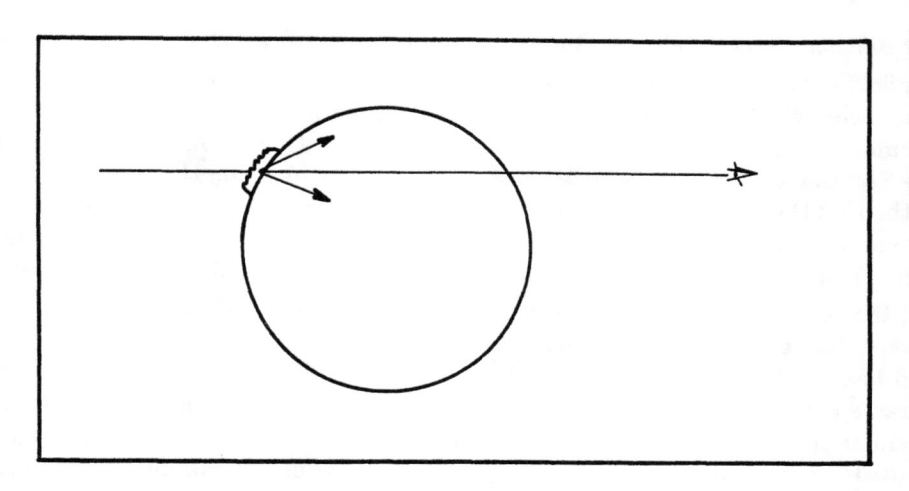

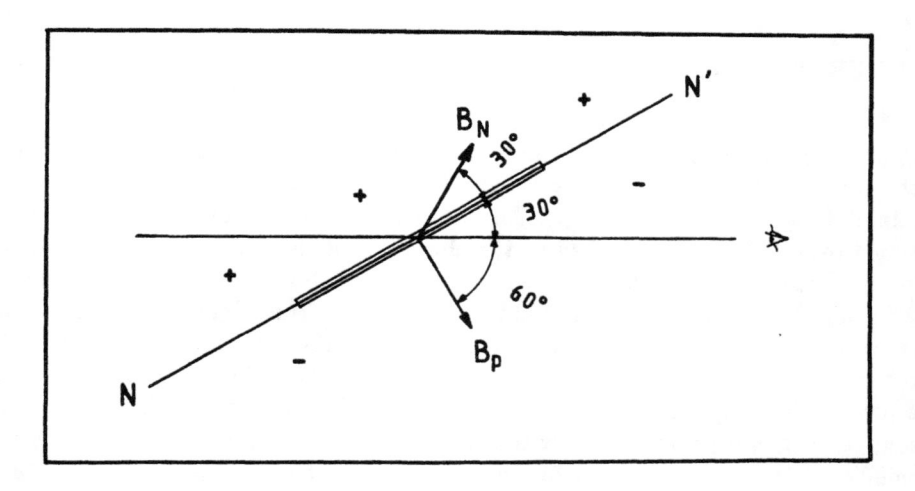

FIGURE 4.3. — Top: Polarimetric diagnostics always yield two possible magnetic field solutions symmetrical about the line of sight. Bottom: Given the most frequent orientation of prominences on the Sun, one generally observes a magnetic field solution B_p with normal polarity having the same topology as a potential field, and another possible solution B_N, with inverse polarity and at a small angle with the prominence long axis.

problem was to wait for some ten years until recent extensive studies on the Hanle effect.

There does not seem to be any way to avoid the 180° ambiguity in Zeeman measurements but there is with Hanle measurements: the degeneracy which provides two indistinguishable solutions depends on the actual geometry of scattering which may be modified, for instance by the displacement of a prominence due to solar rotation (Bommier and Sahal-Bréchot, 1979). An especially interesting situation happens with the observation of optically thick lines, where the prominence's own emission and absorption modify the initial geometry of excitation by a vertical photospheric flux. Recent computations by Landi degl'Innocenti et al. (1987) have shown that observing both an optically thin and an optically thick line should allow a recovery of the magnetic vector $\vec{B}$ without the 180° ambiguity. The trouble is only a matter of experimental accuracy, because scattering of radiation in an optically thick medium results in depolarization, making the quantity to be measured still smaller. A series of measurements obtained at Pic du Midi both in D3 and in Hα are currently being analyzed with this problem in view.

4.2.4 INSTRUMENTAL ACHIEVEMENTS

We have just mentioned the problem of measurement accuracy which reminds us that the polarimetric investigation of prominences entails severe instrumental requirements. As magnetic field diagnostics are based on a polarimetric analysis we need first an instrumental device which introduces only negligible polarization: thus, on-axis instruments are needed and one may introduce tilted mirrors only after the polarization analyzer. At least this is compulsory for linear polarization diagnostics. Zeeman measurements involving circular polarization analysis can possibly withstand oblique reflections although it calls for a careful compensation of the instrumental polarization.

The observation of optically thin emissions from prominences, which is required for Hanle effect diagnostics, means that we observe, above the solar limb, rather faint intensity levels. For several reasons (see Sahal-Bréchot et al., 1977) the He I D3 line is very well suited to the study of the Hanle effect : on average, the equivalent width of D3 is of the order of 2×10^{-4} nm of the photospheric spectrum, nearly 20 times smaller than current Hα prominence emissions. Therefore, the halo of parasitic light scattered by the terrestrial atmosphere, or by an ordinary telescope, is by no means negligible and measurements must be performed with a coronagraph from a high-altitude station (measurements which are performed with classical telescopes and coelostats can be aimed at bright prominences only and, in most cases, they deal with optically thick emissions). Even in the best case the halo of parasitic light is so bright, close to the solar limb, that measurements are impossible for altitudes smaller than 10000 km. Eventually, there is a very troublesome gap between photospheric magnetic measurements and the lowest measurements performed on prominences at the limb.

The last point of interest is the signal to noise ratio: we want to measure a polarization degree of the order of 10^{-2} with an accuracy of say 10^{-2} which means that the noise level must be of the order of 10^{-4} of the intensity. Now, coronagraphs are not very large instruments and if we want a good spatial resolution (some seconds of arc) the integration time can hardly be kept below one minute. Thus, it is very difficult to perform measurements of prominence fine structure as we stress later. One-dimensional detectors have represented a major advance, allowing us to determine the full spectral variation of Stokes parameters across a line. Two-dimensional detectors would add one spatial dimension to the measurement but they have still to be tested for small polarization measurements.

The recent Zeeman measurements obtained by Soviet-French observers were performed with the 53 cm coronagraph of Kislovodsk (with the drawback of one coude-mirror). A complete profile of the V parameter is obtained by scanning a spectral line by means of a Fabry-Perot spectrometer (Nikolsky et al., 1985). The output of the measurement is both the $B_{//}$ component of the field vector and the $v_{//}$ component of the velocity; they are obtained with a spatial resolution of 3 to 6 arc sec.

The French polarimeter designed for the study of the Hanle effect was fed by the 26 cm coronagraph of Pic du Midi (Ratier, 1975). The spectral selectivity was provided by a Lyot filter which isolated at the same temperature the D3 and the Hα lines, in addition to the Hβ line in a final version. The measurement of D3 only was performed along three 10 second integrations. The measurement of D3, then of one Hydrogen line, required altogether 4 minutes because different adjustments of the coronagraph were needed (remember that a coronagraph is an essentially chromatic instrument). Thus, the time constant of measurements for two lines was much worse than those with one line only. The spatial resolution was fixed by a field aperture of 5 arc sec.

The American Stokesmeter was fed by the 40 cm coronagraph of Sacramento Peak. A spectrograph provided a good spectral resolution resolving the D3 line profile (Figure 4.1, bottom). In the first instrument (Stokes I; Baur et al., 1980) the Stokes profiles were obtained by spectral scan, whence a poor time resolution (about 20 minutes for a complete scan, with a spatial resolution of 10 × 4 arc sec.). Then, in the Stokes II (Baur et al., 1981) a 128-element detector allowed the whole Stokes profiles to be obtained in a single exposure, decreasing the time for one measurement down to 1.5 minute, for the same spatial resolution.

Before describing the valuable results which have been available thanks to these different instruments it is worth considering also some indirect approaches which can yield useful information on prominence magnetic fields, without dealing with the polarization properties of spectral lines.

4.3 Indirect magnetic field determinations

Beautiful movies of prominences observed at the Sun's limb have revealed trajectories which are very suggestive of magnetic forces, the role of gravity being certainly not dominant. This is particularly true at the time of a prominence destabilization, when rising material follows curves which look much like lines of force of the coronal magnetic field. Thus, it was quite natural to ask whether it is possible to infer some properties of the magnetic field from the observed motions of prominence knots.

After earlier attempts this method has been recently developed with some success by Ballester and Kleczek (1984): the basic idea is that curved trajectories imply that magnetic energy is sufficient to control the motion of prominence material, and therefore is at least equal to the kinetic energy, whence a relation of the type

$$\frac{1}{2} \rho v^2 \leq \frac{B^2}{2\mu} \qquad (4.5)$$

will provide a lower limit for B if ρ and v are known.

The cinematography of prominence motion provides directly two components of the velocity; although it would be possible to measure simultaneously the third component with the help of a spectrographic device, such complete sets of observations are rare concerning fast-moving objects. In most cases, it has been assumed that one can know the full velocity vector, at different points of the prominence, starting from only two measured components. Given some reasonable hypothesis, it is probably rather safe as has been shown in the past and, more recently, by Ballester and Kleczek (1983) and Ballester (1984). Eventually, one can consider that v is reasonably known and future investigations, with such devices as the "MSDP" (Multichannel Subtractive Double Pass Spectrograph) will improve the situation.

We are left with ρ which may give more trouble since the determination of the density, especially of the local density, is a crucial problem in prominence studies (Section 2.2). Happily, even if ρ is uncertain by one order of magnitude, as one can fear, it enters in the determination of B by its square root only so that the final accuracy on B is not completely spoiled.

By this interesting method, Ballester has been able to derive a list of magnetic field strengths measured in various types of prominence. Obviously, the method can be successful only in prominences which display fast motions, i.e. in active and eruptive prominences. A very interesting feature is that the resolving power can be excellent (down to one arc sec.) so that the averaging effects which badly hamper polarization diagnostics are avoided.

In truly quiet prominences there are almost no motions and even the sign of the vertical velocity has been a point of controversy (Section 2.3). Thus it is almost impossible to derive any field strengths from a kinematic study. It is tempting to search some other guidelines and the shape of prominence fine structure has often been considered. But there is no guarantee that the more or less vertical structures which are often seen in prominences at the limb indicate a pattern of lines of force with the same direction. This simple view is, in any case, in direct contradiction with the actually measured field as we see later.

Another possible way to gain some knowledge on prominence magnetic fields is to study prominence oscillations. A comprehensive description of this method can be found in Tandberg-Hanssen (1974) and it is not useful to repeat here the discussion. There seems to be different types of possible oscillation, for instance vertical or horizontal oscillations which can be studied either on the disc (Malherbe et al., 1987) or at the limb. The phenomenon depends of course on the topology of the magnetic field which supports prominence material: former papers on this topic (Hyder, 1966; Kleczek and Kuperus, 1969) consider the classical magnetic pattern of the Kippenhahn and Schluter (1957) model (but the investigation probably would remain significant for different models as long as one assumes that the prominence material is supported in a magnetic trough). In such conditions, the behaviour of the oscillation depends upon the field strength which, for reasonable values of the other parameters, turns out to be of the order of 10 Gauss. In recent years, a large number of observational investigations has provided new data about prominence oscillation (see the review by Tsubaki, 1988). There seems to be two families of oscillations with a period of 3 - 10 minutes for the short ones, and a period of 40 - 80 minutes for the longer. Theoretical modelling to link these new data with the prominence magnetic field has still not been achieved.

Given the efficiency of computations resulting in 3D - maps of the coronal magnetic field one can ask finally whether such types of investigation are valuable for the study of prominences. Actually, after some interesting work by Rust (1970) there has been almost no attempt to extrapolate the photospheric field into the region occupied by a prominence. It is true that the geometry of the solar magnetic field can change drastically at the chromospheric level ("canopies") and that we are interested in the coronal magnetic structure over neutral lines of the photospheric field, that is over regions where the photospheric field is faintest and most difficult to measure. In the following section we recall that the photospheric field around neutral lines is rather poorly known. A better knowledge of at least the actual structure of photospheric neutral lines is a compulsory step before any major attempt to extrapolate the prominence magnetic field from the photospheric level.

4.4 Magnetic field at the photospheric level

As we have already mentioned it was the work of Babcock and Babcock (1955) which showed up the remarkable association between prominences and the Sun's magnetic field: the authors noticed that prominences, seen on the disc as dark filaments, appeared either near the borders of active regions, preferentially on the high latitude side, or along the neutral line dividing a bipolar active region. Subsequent work (Stepanov, 1958; Howard and Harvey, 1964; Smith and Ramsey, 1967) fully confirmed this important discovery and it was possible to state safely (Martin, 1973) that "the first necessary condition for the existence, and probably the formation, of a filament is that the vertical component of the local magnetic field be oppositely directed on either side of the filament location". The only exceptions to this rule happen sometimes at the center of an active region, where one can observe one filament end leaving the neutral line and pointing directly towards a sunspot (MacIntosh, 1972a); but, in such a case, the filament is rather unstable with conspicuous material streaming towards the sunspot.

The concept of a "neutral line" entails several aspects which are worth remembering here : first, a neutral line is determined by the vanishing of the longitudinal field $B_{//}$. We have already commented on the lack of sensitivity of the transverse field measurements which prevents the measurement directly of those regions where the field vector is zero. Therefore, such a neutral line is, strictly speaking, a region of the photosphere where the field is horizontal, possibly zero. There are good reasons to believe that the transverse field is also small because, for most neutral lines, one still measures $B_{//} = 0$ at large distances from the disc center, where the line of sight is far from the Sun's vertical. But one should always bear in mind that the true magnetic configuration of a neutral line, and particularly the value of the transverse field component, is very poorly known.

The next problem is the change of appearance of a neutral line with increasing spatial resolution: if we wish to study in detail the neutral line environment we need a good magnetographic sensitivity which prevents the use of very small scanning apertures. For instance, the large-scale magnetic structure of the Sun is well determined with the help of the Stanford magnetograms (Duvall et al., 1977) which have a very good sensitivity of 0.05 Gauss but a spatial resolution of only 3 arc min. A detailed study of the magnetic field topology on each side of the neutral line requires a better resolution: reducing the field aperture from 70 arc sec (Babcock and Babcock, 1955) down to 10 arc sec. (Howard and Harvey, 1964) already revealed the very irregular path of a neutral line, particularly in the vicinity of active regions. Rust (1970), working with a 5 arc sec. field aperture, has shown that a neutral line is bounded by irregular patches of opposite polarity (see Figures 1 and 2 in Rust's paper) with measured field strength of the order of 10 Gauss which must be larger if the size of the magnetic elements is smaller than the scanning aperture. The appearance of the neutral line is not the same in active regions amidst relatively strong fields (see for instance Von Kluber, 1967) and at high latitudes where one observes ill-defined neutral bands rather than neutral lines: of course the difference arises partly from the existence of the instrumental threshold but it seems nevertheless that the appearance of prominences is linked in some way to the magnetic field gradient near the neutral line (Maksimov and Ermakova, 1985).

With modern measurements of the very small scale structure of the photospheric magnetic field, including quiet regions (Stenflo, 1973; Zwaan, 1987), the location of neutral lines relative to concentrated flux tubes becomes a question of great interest (especially if one wants to extrapolate the photospheric field, as mentioned in the previous section). There have been few investigations on this important topic even though the work by Giovanelli (1982) helps an understanding of the nature of magnetic fine structure far from active regions. A possible approach for studying the location of neutral lines relative to regions of concentrated flux could be also to clarify the relation (if any) between the path of neutral lines and the supergranulation network (Ploceniak and Rompolt, 1973). As for neutral lines within active regions, the experimental approach is less difficult since the field strength levels are higher. The recent result by Martin (1986) looks very promising since it shows a relation between the appearance of dark material making a filament and the cancellation of small magnetic knots adjacent to the neutral line (among several interpretations the author suggests that the disappearance of opposite-polarity vertical fields could be the signature of a reconnection process yielding horizontal lines of force at the place where a filament begins to condense; see also Zwaan (1987) and Priest (1987)). Obviously our understanding of the prominence phenomenon needs as well as such types of observation the important measurements in the prominence plasma itself.

Bearing in mind that the gradient of magnetic field on each side of a neutral line may be important for the existence of a prominence, we may study separately various quiescent prominences according to the location on the Sun of the underlying neutral line. Tandberg-Hanssen (1974) already introduced the distinction between A-type neutral lines, which separate the two magnetic polarities of a bipolar active region, and B-type neutral lines which are found in between two different active centres. The A-type neutral lines were much more familiar after the classical studies by d'Azambuja and d'Azambuja (1948) and Kiepenheuer (1953). However, the B-type must be considered equally important since it has been shown (Tang, 1987) that it involves a large proportion of prominences. A major difference between both types is obviously that the field strength is much larger around A-type neutral lines.

We believe that one must add a third category, the C- type, which represents the neutral line associated with the polar crown. Although it shares with B neutral lines the property of separating weak field regions, it has several specific features including its particular situation on the polar side of expanding active regions (Bumba and Howard, 1965) and the well known drift towards the poles (Waldmeier, 1957; Hyder, 1965b): clearly enough, the evolution of a C neutral line is linked with global solar magnetism (MacIntosh, 1980; Hansen and Hansen, 1975). Notwithstanding the rather faint magnetic fields which are measured at the photospheric level in high latitude regions, the polar crown displays numerous examples of high, long-lasting, interesting prominences.

It may be that there are some morphological differences between the prominences which are formed over A, B and C neutral lines. For instance, it is well known that the prominences of active regions (A neutral lines), which are conveniently observed on the disc as dark filaments, are almost invisible at the limb because their altitude is never large. Recent studies by Kim et al. (1987) confirm that the maximum height of prominences is not the same for high-latitude and low-latitude objects.

We do not want to dwell on the morphological signature which is observed at the chromospheric level since this is treated in Sections 2.1, 3.4 and 3.6; but we must remember that the appearance of the neutral bands is very different in weak field regions from that found in active regions (Foukal, 1971), where the elongated pattern of fibrils parallel to the neutral line is a strong argument in favour of a chromospheric field making a small angle with the neutral line (Bruzek and Kuperus, 1972). Although the neutral lines can be followed continuously in some instances, passing from A to C types, or from B to C types, it is not certain either that the associated magnetic structure is always the same or that the prominences over A, B or C neutral lines represent exactly the same physical phenomena. We come back later to this important question, after we have described magnetic field measurements concerning the prominence plasma itself.

Although this is beyond the scope of the present section, we want finally to give a few comments about the possible measurements of magnetic fields near prominences, in the coronal medium: it is well known that the polarization of coronal forbidden lines bears the signature of the coronal magnetic field (Charvin, 1965). However, it is not straightforward to derive the parameters of the coronal field; the polarization degree is strongly influenced by collisional processes depending on the local density (Arnaud, 1982b); the polarization direction gives the direction of the magnetic vector projected on the plane of the sky with, however, a possible $90°$ ambiguity which is troublesome (Arnaud, 1982a). However, the most serious drawbacks are the following: first, as measurements are very difficult, the spatial resolution is poor, of the order of one arc min., even in the most favourable case of the Fe XIII λ 10747 Å line (Arnaud and Newkirk, 1987). Next, one must remember that there are coronal cavities around quiescent prominences (Chapter 3) so that the coronal emission measured near a prominence at the limb is likely to arise from foreground or background regions, the line of sight integration effects being very severe for such optically thin emissions. In view of these problems prospects for the investigation of the Hanle effect in coronal permitted lines are promising but still not achieved (Sahal-Bréchot et al., 1986).

4.5 Main features of the magnetic field in quiescent prominences

4.5.1 FIELD STRENGTH

We have gathered in Figure 4.4. five representative histograms of the field strength measured in quiescent prominences by different groups.

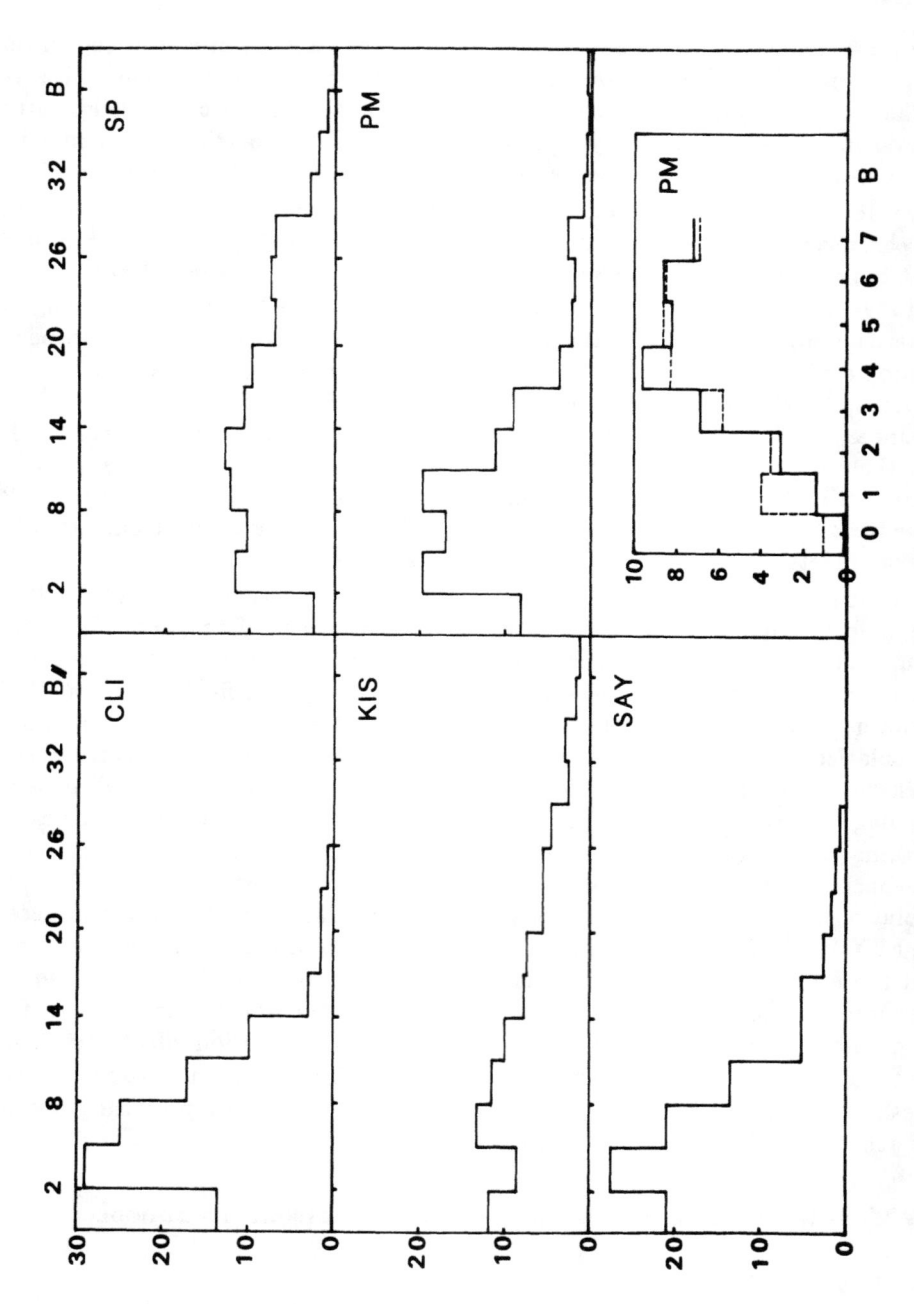

FIGURE 4.4. — Histograms of the field strength measured in quiescent prominences: left, Zeeman measurements from Climax, Kislovodsk and Sayan Observatories; right, Hanle measurements from Sacramento Peak and Pic du Midi.

The left column concerns measurements of the longitudinal component $B_{//}$ deduced from the Zeeman effect. The first one (top) is the well known histogram which summarized the observations obtained with the Climax magnetograph (Tandberg-Hanssen, 1970); it was this instrument which established firmly that the order of magnitude of the field strength is 10 Gauss (0.001 Tesla). The other histograms at the left have been published by two Soviet groups which are presently working on this subject (middle: Kim et al., 1988; bottom: Bashkirtsev and Maschnich, 1987). The right column displays measurements of the total field strength B, as deduced from the Hanle effect analysis, according to the measurements performed at Sacramento Peak (top: this histogram has been actually presented by Nikolsky et al., 1984) and at Pic du Midi (middle: Leroy, 1988).

The agreement between these various histograms is rather good and, summarizing, one can state that 90% of the field strength values are in the range 3 - 30 Gauss. Further, it is of interest to notice that the Zeeman effect and the Hanle effect do give the same result which implies, as we see later, that the field is not extremely inhomogeneous, because averaging effects would be very different in both methods.

It is not easy to say to what extent the agreement on global histograms still holds for individual observations because there has been a very small number of prominences observed with both methods. Bashkirtsev and Maschnich (1987) have presented the interesting case of a polar prominence (1980 August 13) observed at Pic du Midi (Hanle effect) and at the Sayan Observatory (Zeeman effect). The Hanle effect yielded an average field strength of 10.2 Gauss, with the field vector at the angle of 140° with the line of sight, whence the average longitudinal component $B_{//} = -6.7$ Gauss. The Zeeman effect yielded $B_{//} = -4.0$ Gauss. The complete scanning of the four Stokes parameters profiles provided by the "Stokes II" also gave the opportunity to compare the field diagnostics obtained from the Hanle effect (analysis of Q and U profiles) and from a combination of the Hanle and Zeeman effects (analysis of the V profile). The investigation performed by Athay et al. (1983) has revealed that in 11 cases out of 16 the agreement was good.

Returning to Figure 4.4 one can also say that the reasonable agreement between the histograms of B and $B_{//}$ is possible only if $\vec{B}$ is generally not too far from the line of sight, i.e. more or less paralled to the solar equator in most cases. But we know that quiescent prominences are stretched in longitude by differential rotation and are almost parallel to the equator in the case of the polar crown. Thus we have a first indication that the magnetic vector makes a small angle with the prominence long axis, which can be confirmed by more convincing arguments.

Even though many quiescent prominences are stretched in longitude some make a large angle with the line of sight and can even lie almost in the plane of the sky when observed at the limb. In such a case the $B_{//}$ values are found around 0 only because the magnetic vector is perpendicular to the line of sight. It means that the small $B_{//}$ end of the histograms on Figure 4.4, left, is not easy to interpret since it results partly from perspective effects. If we want to know what happens really for very small prominence fields we must rely only on the B measurements provided by the Hanle effect. It is the reason why we have given (Figure 4.4, bottom right) a detailed histogram of the Pic du Midi measurements for small field strength. It is built with the help of a large number of measurements: 1000 measurements on prominences of the polar crown (dotted line) and 1500 measurements on prominences of medium and low latitude (full line); it proves that there are almost no prominences with field strength below 2 Gauss. One possible interpretation could be that there are no regions in the lower corona with a magnetic field smaller than 2 Gauss: we consider this hypothesis unlikely since the histograms for polar regions and low latitudes are almost the same. More probably, the "prominence phenomenon" cannot appear, or endure, for field strengths below 2 Gauss. (Incidentally, if one adopts an electron density of 3×10^{10} cm^{-3} (Bommier et al., 1986b) and a ionization degree HII / HI of 0.3 (Hirayama, 1986) one finds that a field strength of 2 Gauss corresponds to a plasma beta of around unity).

Let us focus now on the tail of the histograms towards strong B or $B_{//}$ values. There, we can see noticeable differences between various results and there are enough grounds to suspect the composition of the different prominence samples. As has been noticed in Sections 1.1 and 2.1, the term "quiescent prominence" is used for rather different objects, the extreme cases being the low filaments of active regions and the high quiet prominences of the polar crown. There is no reason why those objects should have the same magnetic structure and it has been actually reported (Leroy et al., 1984) that there are probably two classes of quiescent prominence with different magnetic properties: the difference lies mainly in the field polarity (section 4.5.4) but is also noticeable in the field strength: roughly speaking, the high-latitude quiescent prominences have a typical field of the order of 8 Gauss while for the low-latitude quiescents near active regions the typical value is instead 20 Gauss. A similar conclusion has been obtained by the Izmiran group (Nikolsky et al., 1984; Kim et al., 1988) who show that the field strength histogram displays significant changes when the data are treated separately for high and low latitudes. Further, it seems (Leroy et al., 1984; Kim et al., 1987) that the typical height is not the same for the different families of prominence, being smaller for active region prominences. Of course, there is still some work to do before adopting definitely the distinction which has just been sketched: for instance one can question about possible intermediate situations (what is the status of prominences over B type neutral lines i.e. amidst weak field regions of low latitude?) One notes also that if the dichotomy is well contrasted it must introduce some cyclic variation in the average magnetic field observed at a given epoch, because prominences of different types are not found in equal proportion along the 11-year Cycle. As a matter of fact, Hirayama (1985) has noticed that the average prominence field is larger at the time of solar maximum, which is well consistent with a larger proportion of active-region prominences at this time.

With the help of the numerous Pic du Midi observations we have searched possible sub-classes concerning the magnetic properties of quiescent prominences and, in particular, we have searched for a possible link between the morphology of polar crown prominences and the measured magnetic field. We suggested (Leroy et al., 1984) that the smallest field strengths are found in very quiet prominences with essentially vertical structures. However, Kim (1987, private communication) has found some examples of vertically structured prominences with measured field strengths up to 30 Gauss and, following her remark, we looked again in our files and we did find some examples of the same kind. Eventually, we come to the surprising conclusion that inside a given class of prominence (e.g. the polar crown) there is no tight relation between the apparent morphology of emitting material and the measured magnetic field: the same prominence can exist for a field strength varying by one order of magnitude.

The preceding conclusion is consistent with the observational fact that the field strength measured in prominences shows almost no dependence on the intensity of monochromatic emission (see Nikolsky et al., 1985, figure 3). This is particularly well seen when the apparent surface of a large prominence is scanned with a magnetograph which does not indicate any particular magnetic signal from the brighter, or from the darker, structures. In several cases, high-latitude quiescent prominences display a very smooth pattern of magnetic field strength which contrasts with the highly structured emissivity (Leroy, 1988). This trend, of course, is partly due to the averaging effect resulting from a poor spatial resolution (some seconds of arc) but a part of the effect is real: in some cases we have found that the standard deviation of the B values, when corrected for the measurement errors, are smaller than the standard deviation of the prominence line intensity. This question is related to the problem of the fine structure of prominence magnetic fields which will be considered in section 4.5.5.

The height dependence of the field strength in quiescent prominences has been a subject of interest for a long time, especially after Anzer (1969) has shown that stable configurations within the Kippenhahn and Schluter model (1957) should correspond to an increase of the field strength with altitude. As a matter of fact, the observations by Rust (1966, 1967) did indicate a trend in favour of an increase of B with altitude at the rate of 1×10^{-4} Gauss km^{-1}. Then, using a large sample of 120 prominences of the polar crown, Leroy et al. (1983) confirmed Rust's result and derived an average gradient of 0.5×10^{-4} Gauss km^{-1} (Figure 4.5, bottom). Later on, Bashkirtsev and Maschnich (1987) also observed a positive vertical gradient of the field strength which is much larger (4×10^{-4} Gauss km^{-1}). However, the H.A.O.-Sac Peak data (Athay et al., 1983) do not show any significant relation between altitude and field strength.

Although these different results look confusing we believe that there is no major contradiction between the various observations in view of the two following points. (1) The vertical increase of B is but a small effect which is easily masked by the

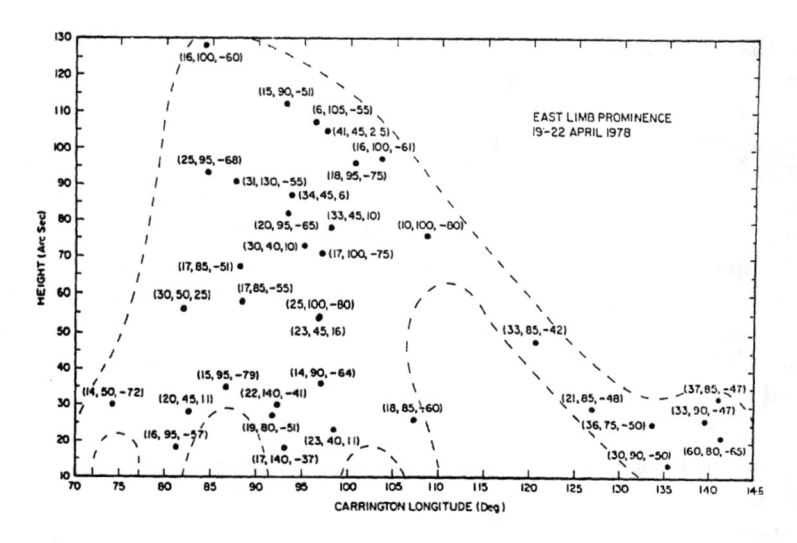

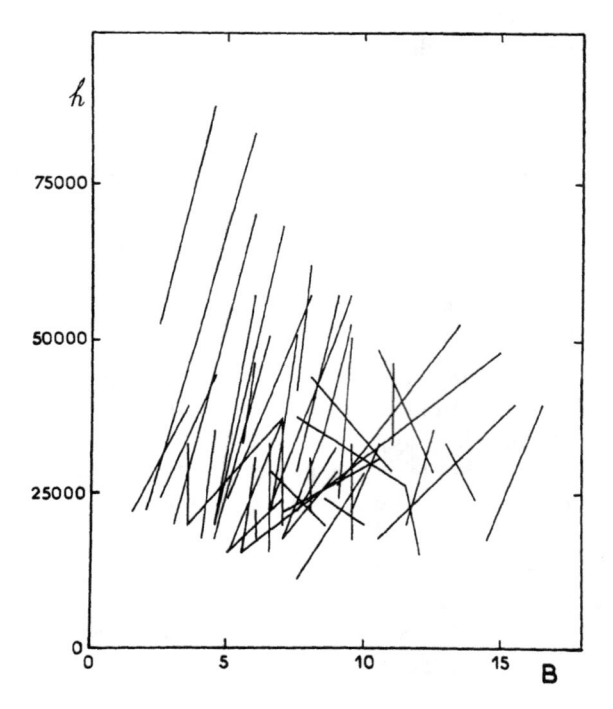

FIGURE 4.5. — Top: a map of the magnetic field parameters (B, ψ, θ) at several points of a prominence (Querfeld et al., 1985). Bottom: average internal variation of B with altitude for a sample of polar prominences (Leroy et al., 1983).

intrinsic prominence "noise". Thus it can be observed only in very quiet objects which have a uniform distribution of B, i.e. essentially in high-latitude prominences. (2) One should not confuse the slow internal increase of B inside high prominences with the difference between average values of B for low (e.g. B = 20 Gauss over active region neutral lines) and high (e.g. B = 8 Gauss in the polar crown) prominences. We guess that the increase of B with altitude, which has been established only for high-latitude, high prominences, might be the result of some relaxation process which can be efficient in long-living objects. Nothing has been proved, up to now, concerning the active-region prominences in which the small vertical extent makes, in any case, the investigation much harder.

4.5.2 ANGLE WITH HORIZONTAL

The $B_{//}$ measurements provided by Zeeman magnetographs were unable to yield any conclusion on the field orientation relative to the vertical. However, in most papers about quiescent prominences until the 80's it was widely agreed that the field configuration is more or less horizontal, as forecasted by most theoretical models. First measurements based on the Hanle effect (Leroy, 1979), obtained with one line only, were sufficient to exclude vertical or strongly inclined solutions (section 4.2.2) which was consistent with the absence of vertical field at the photospheric level. It was the work of the American group, after a polarization analysis of the two components of the D3 line, which proved first that the field vector is indeed very close to the horizontal plane (Athay et al., 1983; Querfeld et al., 1985). According to Athay et al. (1983) the average field vector in quiescent prominences departs by less than 10° from the horizontal, even though local measurements may show somewhat larger departures. Figure 4.5. (top), after Querfeld et al. (1985) gives a picture of B, ψ and θ in a large prominence.

The same type of investigation was repeated with the Pic du Midi instrument through simultaneous measurements of the D3 and Hβ lines (Bommier et al., 1986a); it ended with essentially the same conclusion, namely that the field vector is almost horizontal in quiescent prominences. In addition Bommier et al. (1986a) investigated the influence on the derived field parameters of unresolved "V" - shaped structures such as the possible magnetic troughs which can support prominence material (see Bommier et al., 1986a, Figure 6a). The conclusion is that measurements are consistent with such a pattern but the angle of the "V" relative to the horizontal plane is less than 30°. It is interesting to remark that Hirayama (1979) had forecasted such a behaviour through a comparison of the size of prominence fine structure with the scale height of a 6000 K plasma. It is also obvious that the agreement between the B and the $B_{//}$ values which has been mentioned in the previous paragraph requires a nearly horizontal field pattern.

It must be stressed that this well defined geometry is found only in quiescent prominences: in active prominences (e.g. surges, loops etc...), which are beyond

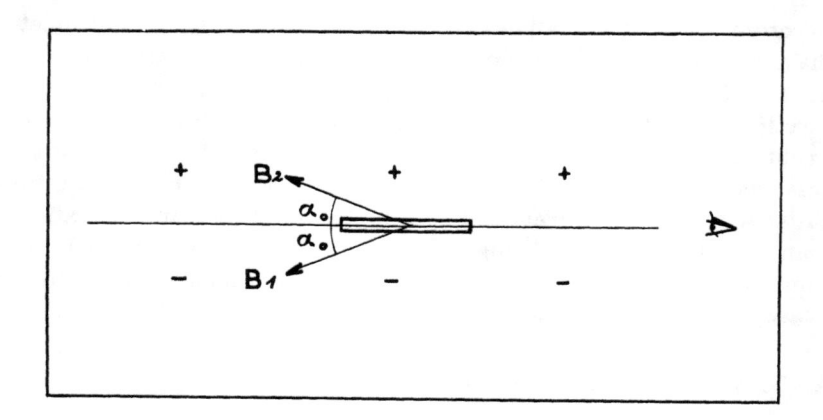

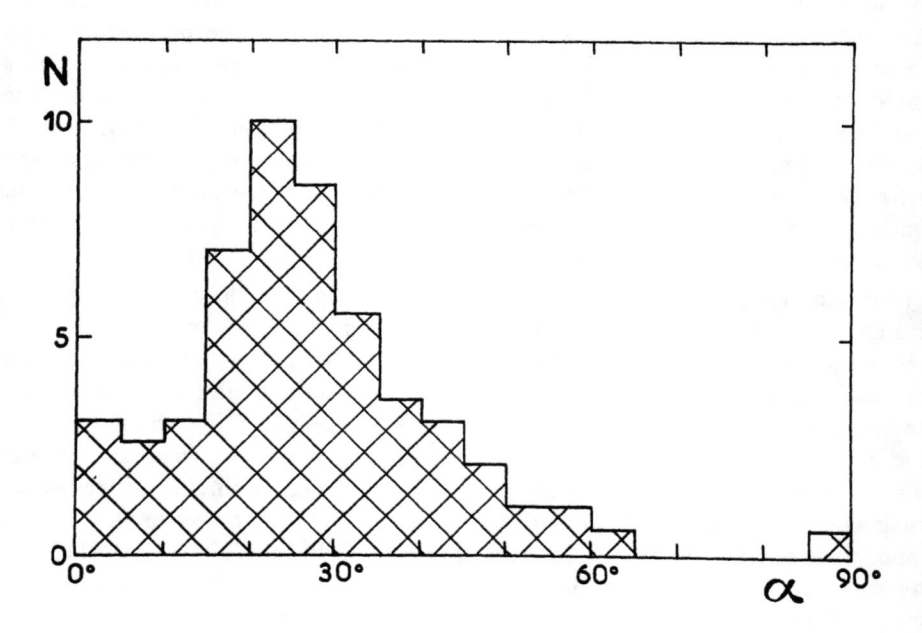

FIGURE 4.6. — Top: Observing a prominence edge-on allows one to determine without any ambiguity the angle α between the field vector and the prominence long axis. Bottom: histogram of the α values obtained for a sample of polar prominences.

CHAPTER 4: OBSERVATION OF PROMINENCE MAGNETIC FIELDS

the scope of this book, one measures field vectors with very different orientations, sometimes nearly vertical (Leroy, 1988). In such cases, it is interesting to note that one observes also fast (say more than 20 kms^{-1}) vertical motions; in quiescent prominences the vertical motions are slow (e.g. < 10 kms^{-1}) as described in Section 2.3. Altogether, the magnetic field diagnostics are well consistent with the observed dynamical behaviour of various prominences (Leroy, 1988).

4.5.3 ANGLE WITH PROMINENCE AXIS

Like the inclination of the magnetic field vector, it was impossible to derive directly the azimuthal orientation of the field vector from $B_{//}$ measurements only. However, taking advantage of the different orientations of prominences relative to the Sun's equator it is possible, through a statistical treatment, to find the most probable angle α between the prominence long axis and the magnetic field vector. This analysis was performed first by Tandberg-Hanssen and Anzer (1970) on a sample of 70 prominences observed at Climax; it proved for the first time that the magnetic field is not perpendicular to the prominence plane, as generally expected, but more probably makes a small angle ($\alpha \simeq 15°$) with the prominence long axis.

Analyzing in a similar way the large sample of observations obtained at Kislovodsk, the Soviet-French group reached a very similar conclusion (Kim et al., 1988) and derived a most probable angle of 25° between the field vector and the prominence long axis. Their conclusion is strengthened by the finding that the velocity vector, which should be paralled to the magnetic field, makes also a most frequent angle of 25° with the prominence long axis.

The determination of the field orientation from individual measurements has been possible only with the help of the Hanle effect. However, the 180° ambiguity which has been described previously (section 4.2.3) still hampers the interpretation. The Pic du Midi data statistically confirm (Leroy et al., 1984) that the field vector makes a small angle with the prominence long axis. Further, there is at least one situation where the 180° ambiguity does not prevent an exact determination of the angle α between the field vector and the prominence long axis : when a prominence is observed edge-on, the two indistinguishable solutions which are symmetrical about the line of sight yield the same α value (Figure 4.6, top). Once a large sample of polar prominences has been observed it becomes possible (Leroy et al., 1983) to obtain the distribution curve of the α angle which is displayed in Figure 4.6 (bottom). Again, one finds clearly a maximum for $\alpha = 25°$.

Altogether we think that there is little doubt about the conclusion that α is around 20° since this figure has come out both from Zeeman studies and from Hanle diagnostics. Nevertheless, some authors (Querfeld et al., 1985) still favour the quasi-potential configuration ($\alpha = 90°$) and further determinations of α are still useful and welcome.

With an α angle of 20° between the field vector and the prominence long axis it is the component of the field parallel to the neutral line which is by far dominant. A very interesting feature reported by Hyder (1965a) and by Rust (1967) is that at the latitude of the polar crown the prominence field orientation remains the same, opposite in both hemispheres, and is linked to the solar cycle. This general organization of the field was confirmed by the early Hanle measurements (Leroy, 1979). Later on, it became clear that the prominence field component parallel to the neutral line rotates by 180° if one passes from the polar crown to the next neutral line at lower latitudes (Leroy et al., 1983). Eventually, the large amount of data gathered at Pic du Midi over 9 years has revealed a clear dependence between the prominence field and the large-scale solar magnetic field as described for instance on the maps of MacIntosh (1972b). Obviously, this magnetic field organization cannot be followed easily at active latitudes but it shows up clearly at medium and high latitudes (Figure 4.7), in close relation with the phase of the Cycle (see Leroy et al., 1984, Figure 13).

Before concluding this section it is worth giving a last comment: many prominence models are presented with a field vector perpendicular to the prominence long axis and it is stated that the magnetostatic equilibrium is not changed by an additional field component along the neutral line. In this view, the component parallel to the neutral line is allowed, but certainly not compulsory. On the other hand, observations show that the field component along the prominence long axis is more important and, from the observer's point of view, it would be much more satisfactory if models include this feature as a starting condition, not as an incidental property (see e.g. Priest et al., 1988).

4.5.4 MAGNETIC STRUCTURE WITH NORMAL OR INVERSE POLARITY

In the book by Tandgerg-Hanssen (1974) a comprehensive discussion was given about several prominence models with normal polarity which are more or less similar to the Kippenhahn-Schluter model (1957): in all the cases, the apparent polarity of the field across the prominence sheet is the same as the polarity of the underlying photospheric field. The same property does not hold in another family of models with inverse polarity, initially proposed by Kuperus and Tandberg-Hanssen (1967) and developed by Kuperus and Raadu (1974): during a discussion held at the time of the Oslo Colloquium (Anzer, 1979) it was clearly shown that the apparent polarity of the magnetic field across quiescent prominences is a good criterion to distinguish the two main families of models which, at that time, were labelled as the Kippenhahn and Schluter (K-S) and the Kuperus and Raadu (K-R) types. Prominence modelling is presented in Chapter 6 and we investigate now the observational aspects only of this important question. Instead of using the more general labels of potential-like (P) or non-potential-like (NP) prominence models we prefer the notation normal (N) and inverse (I) polarity introduced in chapter 1.

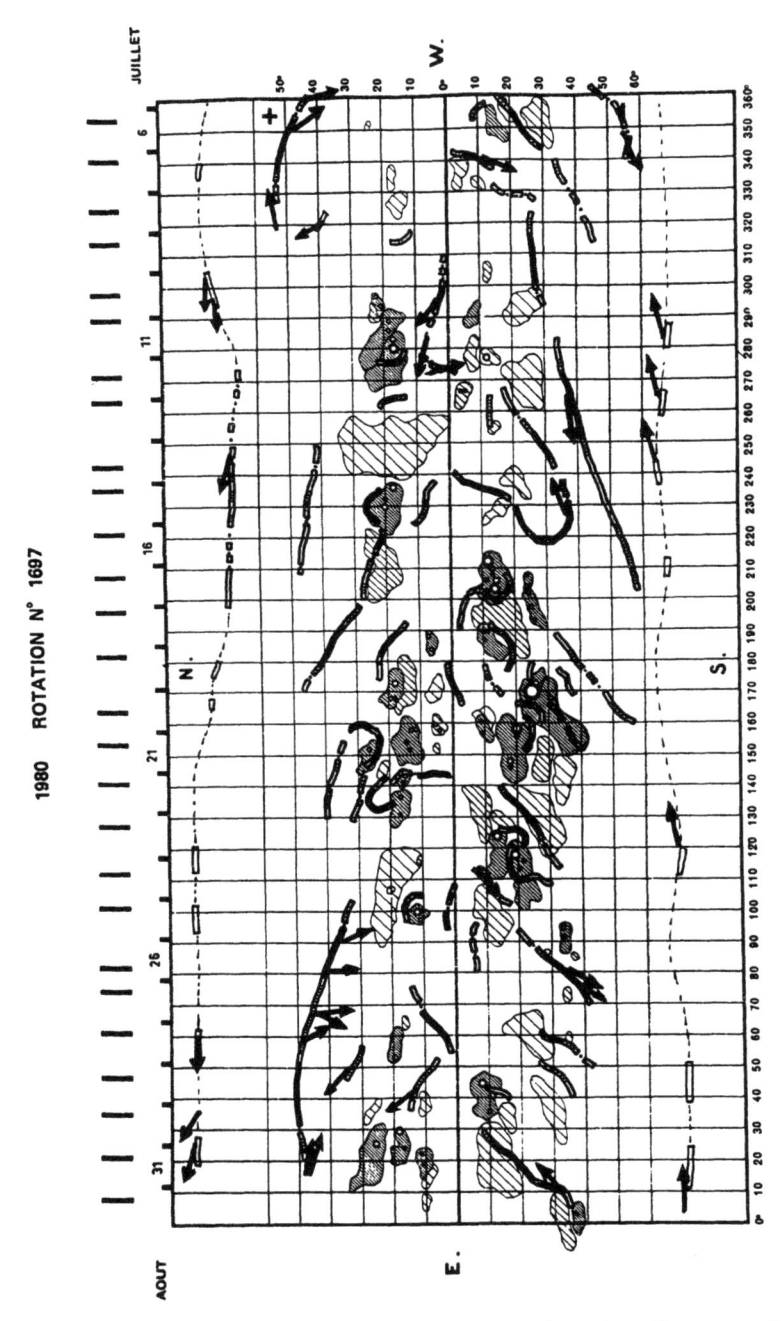

CHAPTER 4: OBSERVATION OF PROMINENCE MAGNETIC FIELDS 103

FIGURE 4.7. — A Synoptic map of the Sun (Meudon Observatory) with the directions of the magnetic vector observed in some prominences displayed as black arrows (Pic du Midi measurements).

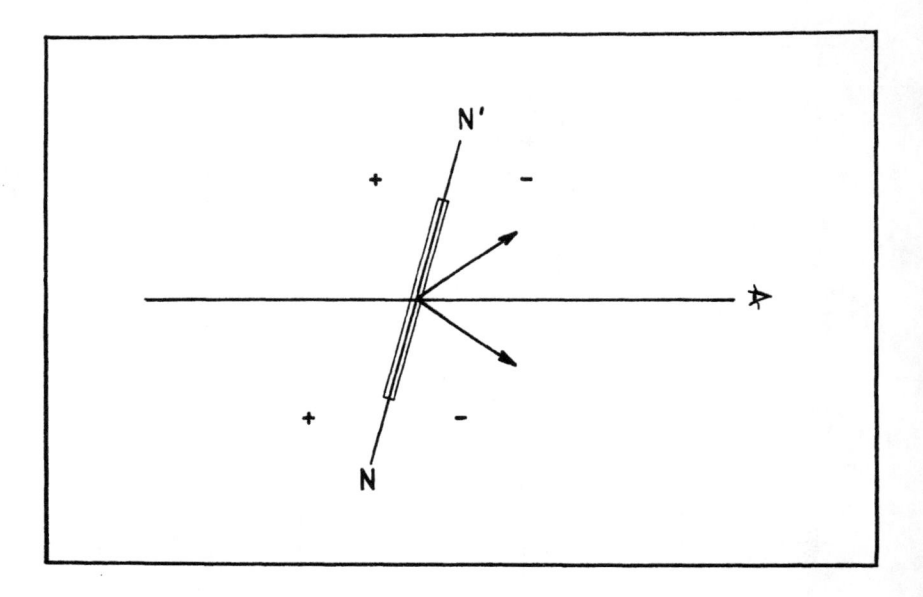

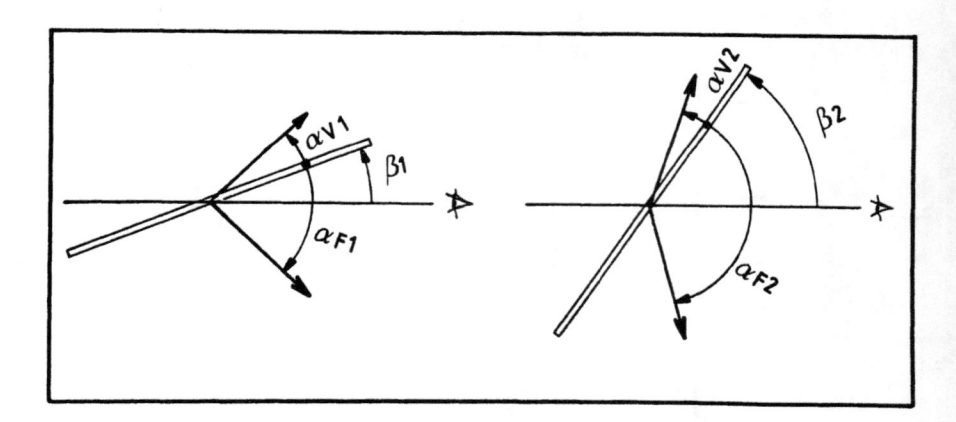

FIGURE 4.8. — Top: Observing a prominence nearly side-on allows one to know without any ambiguity the polarity of the field across the prominence sheet. Bottom: It is possible to select the true field vector, out of the two possible polarimetric solutions, if a given class of prominence can be observed at different angles.

Concerning the apparent polarity of the magnetic field across prominences one must introduce two preliminary remarks: since the field vector is nearly **along** the prominence long axis, it is difficult to determine the small transverse component and, given the unavoidable observational errors (in particular the uncertainties on the prominence geometry and the neutral line location, near the limb!) the solution of the problem is not straightforward. But the most troublesome circumstance arises from the 180° ambiguity on magnetic field measurements that we have already mentioned (section 4.2.3). Although Rust (1979) stressed again at the time of the Oslo Colloquium that all magnetic Zeeman measurements are consistent with normal polarity, we have already explained (Figure 4.3, bottom) that an inverse polarity configuration is equally consistent with the observations: one polarimetric diagnostic yields two indistinguishable solutions which, in the general case, happen as sketched in Figure 4.3, one being normal and the other inverse.

Happily, the Hanle effect analysis provides at least two means to disentangle this dilemma. They have been described by Leroy et al. (1984) and we recall now only their essential features.

In some cases, prominences can be observed nearly perpendicular to the line of sight. In such a geometry (Figure 4.8, top) the two indistinguishable solutions yield the same polarity for the magnetic field across the prominence and therefore we have no doubt about the normal or inverse polarity of the prominence configuration. The only trouble with this method is that it does not work for some types of prominence (e.g. the polar crown) which are never perpendicular to the line of sight. The advantage is that the result is an individual one, some prominences being definitely labelled into the N or the I class.

The investigation by Leroy et al. (1984) started with a sample of 250 prominences of medium and low latitudes observed at Pic du Midi. After cutting the sample down to 120 objects, to discard all those prominences which had not been observed in good conditions (i.e. with a poorly known geometry) it has been possible to retain safely 8 prominences which were of the I type and 5 prominences of the N type (see Leroy et al., 1984, Table 1 and Figures 10 and 12). Thus, for the first time, the existence of I prominences was established on an individual basis; of course, it looked quite necessary to confirm this result with another set of data. The full vector determination in the sample of 14 prominences observed at Sacramento Peak provided 2 cases of I prominences (Athay et al., Table III) which, although they were not considered as very conclusive by the authors of the paper, fit closely the conclusion derived from the Pic du Midi data.

The second way to find out the polarity of prominence magnetic fields is a statistical approach which takes advantage (as in former Zeeman studies) of the various orientations of prominences on the Sun.

Let us consider (Figure 4.8, bottom) a prominence which makes the angle β_1 relative to the line of sight. The true field vector which makes the angle α_{V1} with the prominence long axis is indistinguishable from the symmetrical vector at the angle α_{F1}. Now, if we observe another prominence of the same type under a different geometrical point of view ($\beta_2 \neq \beta_1$) we can assume that the true field vector makes the same angle $\alpha_{V2} = \alpha_{V1}$ with the prominence axis; however, the fictitious symmetrical vector makes the angle α_{F2} and it is obvious that $\alpha_{F2} \neq \alpha_{F1}$. Therefore, with the help of two observations of a given type of prominence it is possible to select the true field vector. As one can easily understand, the presence of errors in measurements and in the knowledge of the geometry makes one prefer a more general statistical investigation which has been described elsewhere (Leroy et al., 1984): among the already quoted sample of 120 medium latitude prominences one finds a clear dominance of I magnetic solutions with an α angle of -25° (the sign - means that we deal with the I configuration).

Again, it was essential to confirm this finding with a different observational basis: the same analysis applied to the Sacramento Peak observations did yield exactly the same conclusion, as was shown by Bommier et al. (1985).

In conclusion, two different ways of investigation (individual and statistical) with two different sets of observations (from Pic du Midi and from Sacramento Peak) have proved that I prominences are present in large number among the population of quiescent prominences. Nevertheless, some cases of N prominences have been identified and one has to deal with at least two different magnetic topologies. In a first attempt to understand the specific features of both families the Pic du Midi data were divided in subsets corresponding to different brightness, height etc... It was found (Leroy et al., 1984) that the N type prominences are essentially low prominences (maximum height smaller than 30000 Km) present at low or medium latitudes; therefore, they must be found mostly over the A-type neutral lines that we have defined earlier (section 4.4). Also, further investigations on the polar crown (C-type neutral line) proved that all these prominences display the I polarity. We have already reported (section 4.5.1) that the N and I prominences have a different characteristic field strength.

At the present time it is very difficult to say whether the two classes of N and I prominences are clearly separated or if there is a progressive shift from one class to the other: remember that possible errors on the geometry (prominences are observed at the limb but the neutral line orientation is determined after disc observations, several days apart) are certainly not negligible given the small value of the α angle between the field vector and the prominence long axis ($\alpha \simeq -20°$ for I prominences and $\alpha \simeq +20°$ for N prominences). The association of N and I classes with specific neutral lines on the Sun (N prominences over A-type neutral lines and I prominences over C-type neutral line of the polar crown) is only a first step which leaves many questions open (e.g. what is the magnetic structure over B-type neutral lines). According to Pic du Midi observations (Leroy et al., 1984) one never observes at the same spot over a neutral line an I prominence overlying an N prominence. But one would like to know whether it exists at some point on a (continuous) neutral line, running from low towards high latitudes, where the magnetic structure changes from one type to the other...

Another point of interest is the relative proportion of N and I prominences at the surface of the Sun: in the Pic du Midi sample the proportion looks to be of the order of 3/4 (I) versus 1/4 (N). However, one cannot take these figures at their face value: remember that N prominences are lower than 30000 km; remember on the other hand (section 4.2.4) that coronagraph observations miss prominences lower than 10000 km because they are masked by too much scattered light. Notice also that most studies on the distribution of prominence heights (e.g. Billings and Kober, 1957) have shown that low prominences are much more numerous than high prominences: therefore the sample of data recorded by a coronagraph is certainly not representative of the actual population of prominences on the Sun. Eventually, the distribution of magnetic structures which have been observed is at least strongly biased in favour of high prominences, that is in favour of the I type. Further studies on the distribution of prominence heights, like that recently achieved by Kim et al. (1987), are needed and could improve our understanding of this question. We would like also to suggest that it might be possible to seek a possible relation between N and I prominences and the different appearances of the neutral line at the chromospheric level (see section 4.4).

4.5.5. HOMOGENEITY OF THE FIELD

We have already reported (section 4.2.4) that the magnetographs which have been designed in the last ten years to measure prominence magnetic fields have an angular resolution of several seconds of arc. Now, it is well known (Engvold, 1976) that prominences display very fine structures, down to the best instrumental resolution of 0.5 arc second. Thus, it is clear that measurements actually mix together the polarized light emitted by many substructures, which is a subject of great concern if the magnetic structure is as inhomogeneous as the monochromatic emissivity. One recalls that, in the case of photospheric studies, the existence of concentrated flux tubes was mis-interpreted for a long time, owing to a lack of spatial resolution!

A general discussion on possible prominence magnetic fine structure has been given by Leroy (1988) and we reproduce now the main arguments of that paper:

(1) – In prominence studies one cannot use the same approach as in photospheric magnetic field measurements i.e. analyzing simultaneously several lines with different Zeeman splitting to circumvent the lack of resolution. However, the Climax magnetograph did measure prominence fields with the lines of Hydrogen, Helium, Sodium and Magnesium and recorded essentially the same field strength. Even though these different emissions originate partly from the same prominence plasma, it is a first indication (Tandberg-Hanssen, 1970) in favour of a rather homogeneous magnetic field.

(2) – A second argument in the same direction comes from the good agreement between the Zeeman measurements and the Hanle measurements which have been

already quoted in section 4.5.1. Such an agreement would be almost impossible if the magnetic field is very inhomogeneous because the averaging processes, when strong and small fields are superposed in one single measurement, are very different for the Zeeman and for the Hanle effect. The Hanle effect is essentially a non-linear phenomenon: if one looks at polarization diagrams more detailed than that of Figure 4.2., one notices for instance that significant changes of polarization direction (i.e. $\varphi \neq 0$ in the frame of Figure 4.2.b) can occur only for intermediate field strength. Small as well as strong horizontal fields can give small φ values only, while important φ values are often observed. On a more general basis, it can be shown (Leroy, 1985) that any strong inhomogeneity of the field strength or orientation involves a decrease of the observed φ values. However the observations have revealed very significant departures from the situation $\varphi = 0$ which must be explained by a rather homogeneous magnetic structure.

(3) – A third type of argument comes from the time evolution of the measured magnetic field : it was shown already with the H.A.O. magnetograph (Harvey, 1969; Tandberg-Hanssen and Malville, 1974) that the time variation of the magnetic field measured in quiescent prominences is rather slow. The same behaviour has been found in the measurements interpreted via the Hanle effect and Leroy (1988) has described the conflicting appearance of rapidly evolving fine structures as seen in Hα light (typical lifetime of 8 minutes according to Engvold, 1976), while the polarimeter provided a rather constant signal. Of course, we are still faced with the averaging effect of too large a field aperture but it seems unlikely that the stable magnetic signal which is measured can be only the random residue of many independent, fast-changing, magnetic substructures. Rather, we guess that there is some kind of large-scale, stable magnetic structure which contains the prominence.

Obviously, the preceding arguments, favouring a large-scale prominence magnetic field, are in contradiction with the instinctive feeling of any prominence observer who is convinced, by his eyes, that the very nature of prominences is inhomogeneous. A simple theoretical answer to this dilemma is that in a low-beta environment very small variations in magnetic field can produce large charges in plasma pressure.

Finally, it is also worth remembering that the observation of a rather smooth magnetic field in prominences does not prove that the magnetic structure is simple: prominence cool material is possibly in equilibrium only in specific parts of magnetic tubes of force (for instance at the bottom of a magnetic trough); therefore it may trace a small part only of the magnetic pattern of the lower corona. We must also stress that the characteristics which have been described in this section concern only quiescent prominences: in active prominences, and also in activated quiescent prominences, there are very clear indications of an inhomogeneous magnetic structure (we have suggested (Leroy, 1988) that the main magnetic singularity of dynamically active prominences is a large scatter of field parameters, rather than a higher level of field strength).

4.6 Some important problems

4.6.1 MAGNETIC FIELD IN SUB ARC SECOND STRUCTURES

Since we have just described the unsatisfactory knowledge of prominence magnetic fine structure it is useful to consider the progress which could be gained with instrumental improvements. We have mentioned in section 4.2.4 some instruments which have provided important results during the last ten years and the question is now whether substantial progress can be achieved in improving their spatial resolution. Since structures with a size as small as 0.5 arc sec. are known to exist in quiescent prominences we would like to perform measurements with a field aperture of 0.5 x 0.5 arc sec. The "Stokes II" (Baur et al., 1981) which was the most powerful magnetograph aimed at prominence studies (owing to its 128-element detector providing a full spectral profile in one exposure) needed an integration time of 1.5 to 2 minutes (Athay et al., 1983) for a scanning aperture of 10 × 3.8 arc sec.: with the same instrument the integration time should increase up to 4 hours if we reduce the field aperture down to 0.5 × 0.5 arc sec. Using a 2D-detector like a CCD, instead of a linear detector, will make it possible to measure the magnetic field in many points along the entrance slit simultaneously. However, the integration time for one measurement remains a matter of hours: given the limited atmospheric seeing, it means that we have almost no chance to reach the expected 0.5 arc sec. resolution. And, last but not least, as the lifetime of prominence fine structures is only a few minutes we have not solved the problem of measuring magnetic field in individual fine structures!

We expect that the direct measurement of magnetic field parameters within prominence fine structures requires an improvement by at least 20 times of the best present instrumental capabilities: it is not clear how and when such progress can be achieved and this is the reason why we believe that indirect magnetic field diagnostics bearing on the fine structure could be the most promising approach to develop in the next few years (section 4.3).

4.6.2 THE PARADOX OF FINE VERTICAL STRUCTURES

Not only are quiescent prominences made of fine structures but these structures often display a distinct pattern : for instance, in prominences of active regions one can observe elongated structures parallel to the prominence long axis and very noticeable motions are generally observed especially on the disc. The high-latitude quiescent prominences observed at the limb show a very different picture with essentially curtain-like vertical structures; slow descending motions are often observed (Engvold, 1976) but they may not represent real motions if we take into account the ascending Doppler motions which are observed on the disc (see chapter 2). Anyway, we meet an apparent contradiction when we compare these vertical fine structures with the horizontal orientation of the magnetic field (section 4.5.2).

A first question is connected with the lack of spatial resolution that we have considered in the preceding section: can we expect that it is the poor spatial resolution of magnetic field measurements which yields a fictitious horizontal orientation for the field vector? The answer is probably no: a vertical magnetic field would yield no magnetic depolarization, which is not consistent, whatever the spatial resolution, with the observations. We can also discard unresolved fine structures like helices with a vertical axis, which could not yield the observed parameters of the polarized light emitted by prominences (Leroy, 1988).

When well visible vertical motions ($v > 20$ kms^{-1}) are observed in active prominences (e.g. surges, loops etc...), one does measure a nearly vertical magnetic field. In quiescent prominences there are only small ($v < 10$ kms^{-1}) vertical motions and we find a horizontal magnetic field pattern. There does not seem to be any major disagreement between the velocity and the magnetic field diagnostics and we must probably agree that vertical fine structures in quiescent prominences do not mark the path of magnetic lines of force.

The same paradox is met with so-called "prominence feet": as we have already explained, the polarimetric measurements bearing on prominences at the limb cannot be performed down to the photosphere but they are limited to the altitudes larger than 10000 km. At such a level the magnetic field measured over prominence feet is almost the same as in other regions of the same prominence (see Figure 4.5., top; see also Athay et al., 1983 and Leroy et al.,1984) i.e. horizontal. Therefore, the often-sketched picture of a bunch of lines of force emerging from the Sun at the place of prominence feet is not consistent with the measurements; further, one should also remember that, at the photospheric level, the neutral line is a region with zero vertical field which contradicts the intuitive explanation of prominence feet (let us recognize, however, that a more accurate comparison of the position of prominence feet relative to the neutral line would be certainly valuable). With this contradiction in mind we can only say that the nature and the magnetic structure of prominence feet is presently almost unknown.

There is a last comment to give concerning vertical structures in quiescent prominences: even though the vertical motions which are measured are very small, they are in any case much larger than the possible plasma velocity across horizontal lines of force (Engvold, private communication). Thus, any vertical motion must probably be understood as resulting from vertical displacements of the whole magnetic structure which supports a prominence.

4.6.3. THE DETERMINATION OF CURRENTS

Modelling a prominence generally entails the conceptual introduction of currents flowing in the prominence and in the corona: these currents are associated with

CHAPTER 4: OBSERVATION OF PROMINENCE MAGNETIC FIELDS

the deformation of an initially potential photospheric field, which provides the support against gravity. Now, even the order of magnitude of these currents is not safely known (see the review by Malville (1979b) for the particular case of active prominences) and one can think that a better determination would be a significant help for the computation of realistic models.

From the observer's point of view the direct approach for deriving the intensity of currents is Ampere's law previously written (equation 1.5) as:

$$j = \nabla \times B/\mu$$

If we are able to determine the spatial derivatives of B we can get j. Let us state that, at the present time, this is beyond observational capabilities, at least if we want to determine j for individual objects: the accuracy on the determination of B is certainly not sufficient to allow the computation of derivatives with a reasonable uncertainty. And, given the problem of line of sight integration, even the spatial coordinates of measurements are poorly known. The only hope to reduce these serious drawbacks is to sum up observations of a large number of prominences so as to derive average variations for the different components of B. The approach would be safer if one could use both limb and disc observations, which could be possible in the forthcoming years (both through Zeeman and Hanle diagnostics). Clearly, indirect methods such as those described by Malville (1979a) have a great interest for this problem.

4.6.4. THE EVOLUTION OF PROMINENCE MAGNETIC STRUCTURE

Since we have found two main types of quiescent prominence with N and I magnetic configurations we may ask whether one can pass from one class to the other through a continuous evolution or if they represent two permanently different physical objects. This problem is linked to the question of prominence life-times that we want to present briefly: although the individual lifetime of prominence fine structure is only several minutes, the life of a prominence considered as a whole is a matter of weeks. Such is also the time-constant found in magnetic field measurements. However, there is a third, longer, characteristic life-time which is that of a neutral line: for instance the neutral line of the polar crown lasts about 5 years, from the Cycle minimum to its disappearance at the pole at the next maximum. It is conceivable that the magnetic structure which is most easily associated with a neutral line is of N type over young neutral lines and of I type later on. As prominences cover only a part of a neutral line there could be also intermediate magnetic configurations at places where no prominences occur.

In his comprehensive review on solar prominences Hirayama (1985) has sketched two possible types of magnetic evolution which are reproduced on Figure 4.9. The evolution "A" (top) fits a model proposed by Pneuman (1983) while the evolution

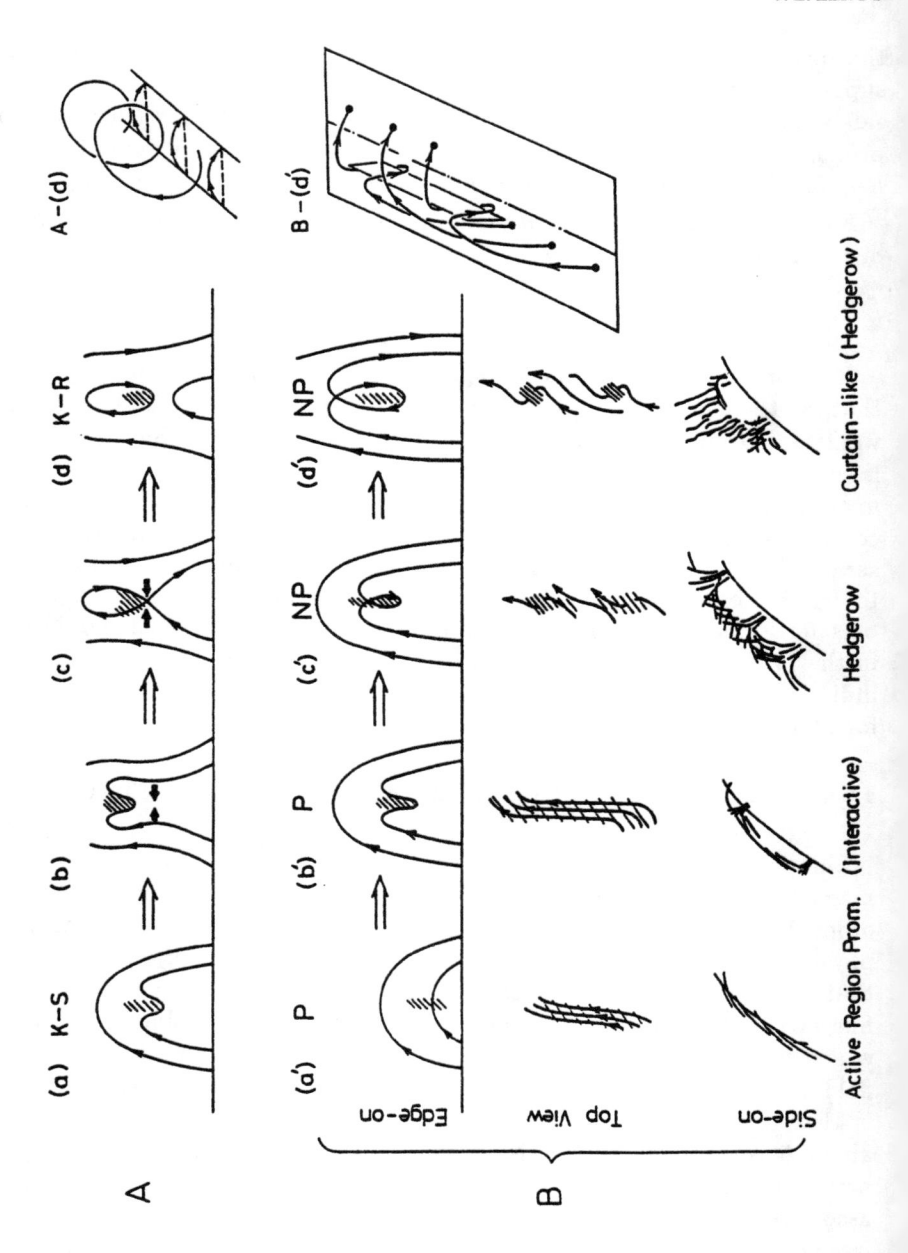

FIGURE 4.9. — Possible evolutions of the magnetic structure associated with a quiescent prominence according to Hirayama (1985). K-S and P represent structures with normal (N) polarity, while K-R and NP refer to those with inverse (I) polarity.

"B" is due to Hirayama himself. He proposes that the I topology can be reached either by reconnection ("A") or by shearing and twisting only ("B"); but, with cool prominence material concentrated at the bottom of magnetic lines of force, the **apparent** magnetic field is almost the same in both cases ...

The sketches by Hirayama are an attempt to synthesize our present knowledge on quiescent prominence magnetic structure; comparing this situation with that prevailing 10 years earlier, it is seen that he is seeking an evolutionary view of the phenomena. One can hope, of course, that the next 10 years will provide still more important advances allowing us to answer the important questions which have been raised in this section.

CHAPTER 5

THE FORMATION OF SOLAR PROMINENCES

J.-M. MALHERBE

Département d'Astronomie Solaire et Planétaire
Observatoire de Paris - Section de Meudon
F - 92195 Meudon Principal Cédex
FRANCE

Recent progress in the understanding of the formation of quiescent solar prominences is summarized in this chapter, from both an observational and theoretical point of view.
It is now well known that the mass of a quiescent prominence (seen in emission at the solar limb) or a filament (generally seen in absorption above the disk) is an appreciable part of the mass of the entire corona (roughly one tenth or more), which makes it difficult to form these structures by coronal condensation alone. Hence, possible mechanisms proposed recently to account for their formation are divided into two categories reviewed below, namely *injection* (of the chromospheric material into the corona by siphon flows) and *condensation* (of the coronal plasma itself).

5.1 Introduction

In the first text book on prominences by Tandberg-Hanssen (1974), chapter 4 presents early models which were produced to account for the formation of these structures, by means of thermal instabilities (Field, 1965) and siphon flows. It is now possible to find in the literature several recent review papers concerning prominence formation, as well as proceedings of conferences containing many contributions about this topic (Jensen et al, 1979; and more recently, Poland, 1986, and Ballester and Priest, 1988). Also, an interesting overview of modern observations, which are useful in the study of prominence and filament formation, has been written by Hirayama (1985). As for the basic concepts of the MHD theory of prominence formation and equilibrium, Priest (1982) has described recent models in chapter 11 of his Solar MHD book.

Prominences are structures of the solar corona; but it is commonly thought that they are 10 or 100 times denser and and 100 times cooler (chromospheric-like conditions). They are anchored in the photosphere by feet. The MHD mechanisms which operate in the formation, equilibrium and instabilities of prominences are complex and involve nonlinear effects due to interactions between heating, conduction, radiation, magnetic field reconnection and gravity, which make these beautiful features of the Sun both fascinating and difficult to study.

The contents of the next sections are as follows:

- Recent observations of filament formation and time-scales.

- Basic MHD instabilities; thermal and resistive instabilities.

- Steady state magnetic reconnection of sheared fields and current sheets.

- Static models: thermal instabilities of the coronal plasma in magnetic fields (with and without conduction) in different geometries; condensation modes in cylindrical plasmas (loops) and in magnetic arcades, incorporating the effect of the shear.

- Dynamic models: I - Injection processes: they can be subdivided into surge-like and evaporation-like models. Models suggesting that material is ballistically launched from the chromosphere to the corona at the speed of spicules will be presented, as well as those showing that a sustained heat release in a loop may give rise to a solar evaporation and thermal instability at the top.

- Dynamic models: II - Coronal condensation: recent numerical models solving the 2D resistive-radiative MHD equations in line-tied coronal current sheets will be described; they give evidence of different kinds of condensation (Kuperus-Raadu - KR - and Kippenhahn-Schlüter - KS -, or models of prominences with inverse or normal polarity). In the latter case, a shock condensation mechanism has been recently discovered.

5.2 Overview of Observations

Quiescent prominences (seen at the limb) and filaments (on the disk) appear as thin sheets of dense (electron densities of $10^{9.5}$ to $10^{11.5}$ cm^{-3} according to various authors) and cold material (about 8000 K) suspended in the corona (about 10^9 cm^{-3} and 10^6 K), and linked to the solar surface by feet. As the pressure scale height (about 300 km) is negligible compared to their vertical extension (5000 to 50000 km), the magnetic fields are believed to support material against gravity (with a field intensity of the order of a few Gauss). It is well established that filaments form above "neutral" ($B_{//} = 0$) lines (the filament channel in the photosphere) which separate regions of opposite polarity.

The problem of the hydrogen density and ionization degree (between 0.05 and 1) is crucial for the understanding of prominences: according to the electron density values bigger than 10^{11} cm^{-3} and small ionization degrees found by Landman (1984), the plasma β is about 10 (with B = 3 G) and material is dominated by pressure forces; while, according to measurements by Leroy et al (1984, see also Chapter 4 of this volume), and Bommier (1987), giving densities of 10^{10} cm^{-3}, one finds $\beta = 0.01$ with an ionization degree of 1, and the plasma is dominated by magnetic forces. In the former case, the mass of a quiescent prominence may be comparable to the mass of the entire corona, which makes questionnable condensation mechanisms of the coronal medium, and material would, in that case, need to originate from the chromosphere.

Plage filaments differ from quiescent ones by their size (5000 km high) and magnetic field (10 times larger); they belong to active regions and definitely seem to be magnetically structured. Motions comparable to the sound speed are frequent in these features, and their formation appears to be more likely caused by chromospheric siphon mechanisms rather than by coronal condensations.

Time-scales involved in prominence formation are different according to the size of the structures considered: the overall structure needs between a few days and a few weeks to appear completely, and it is observed to form section by section, foot by foot (Figures 5.1 to 5.3). The time-scale is also a function of the latitude: it increases towards the poles. But in most cases, it is at least 10 times longer than the coronal radiative time-scale (a few hours), which shows that prominence formation involves processes likely to be more complicated than the ideal thermal instability of coronal plasma.

Figure 5.3 shows that one must be extremely careful with the interpretation of filament appearances in terms of formation of new structures: the "Apparition Brusque" (or AP) phenomenon corresponds to the cooling of an existing filament which was previously heated by activity (it involves a time-scale of the order of the radiative time-scale of the corona, about a few hours, which suggests a radiative cooling and condensation of the hot plasma). This difference between time-scales of pure formation and AP has to be kept in mind. A beautiful example showing the cooling from coronal to chromospheric temperatures has been reported by Mouradian et al (1980).

Martin (1986), from recent observations, shows that filament formation is preceded by a continuous accumulation of dense threads (fibrils) along the channel between large-scale opposite magnetic polarities (this is indeed well visible on Figures 5.1, 5.2 and 5.3), and by small-scale magnetic field cancellation. Feet seem to appear first; they are suspected (the result needs to be confirmed) to form (Rompolt, 1986) at the boundary between supergranulation cells (a region of converging flows) and are separated by about 30000 km (Figures 5.4 to 5.8). But Figure 5.8 shows that the whole filament may have formed even if, in $H\alpha$, only feet are visible above the chromosphere. While the magnetic topology and the velocity field at the feet (which form on both sides of a magnetic inversion line) are still a puzzle (non-resolved fields smaller respectively than 10 Gauss and 5 km/s have been observed), Kuperus-Raadu models (1973, for inverse polarity) or Kippenhahn-Schlüter models (1957, for normal polarity) may apply to the overall prominence support.

The fine structure (less than 0.5 arc seconds) seems to exhibit time-scales of the order of 10 minutes only: the strands are probably highly dynamic (Figures 5.5 and 5.7).

But this paper only describes mechanisms recently proposed to account for the formation of the overall structure, since the physics of the fine structure and of the feet is still observationally almost unknown. Rompolt (1986) suggests that filaments may form either in regions of parallel converging mass motions, or anti-parallel converging flows. In the former case, fibrils become symmetrically elongated along the filament channel, while in the latter one, threads overlapping the channel become sheared (Figures 5.1 to 5.3). Converging (or diverging) photospheric motions below prominences are also necessary, from a theoretical point of view, to explain steady upflows, observed by Schmieder et al (1984, and Chapter 2 of this volume) and Engvold et al (1985), travelling across filaments, if one takes the Kuperus-Raadu (or, respectively, the Kippenhahn- Schlüter) magnetic overall structure. Observed converging motions below filaments and upflows in filaments are in agreement with Kuperus-Raadu magnetic configurations, which seem to be the most common on the Sun (Leroy et al, 1984, and Chapter 4 of this volume).

The shear of the magnetic field in prominences, which is emphasized by Leroy in this book, is suggested by most observers to be a basic property of prominences: dense material appears generally in sheared magnetic fields; and quiescent prominences exhibit a larger shear than plage or active-region ones.

118

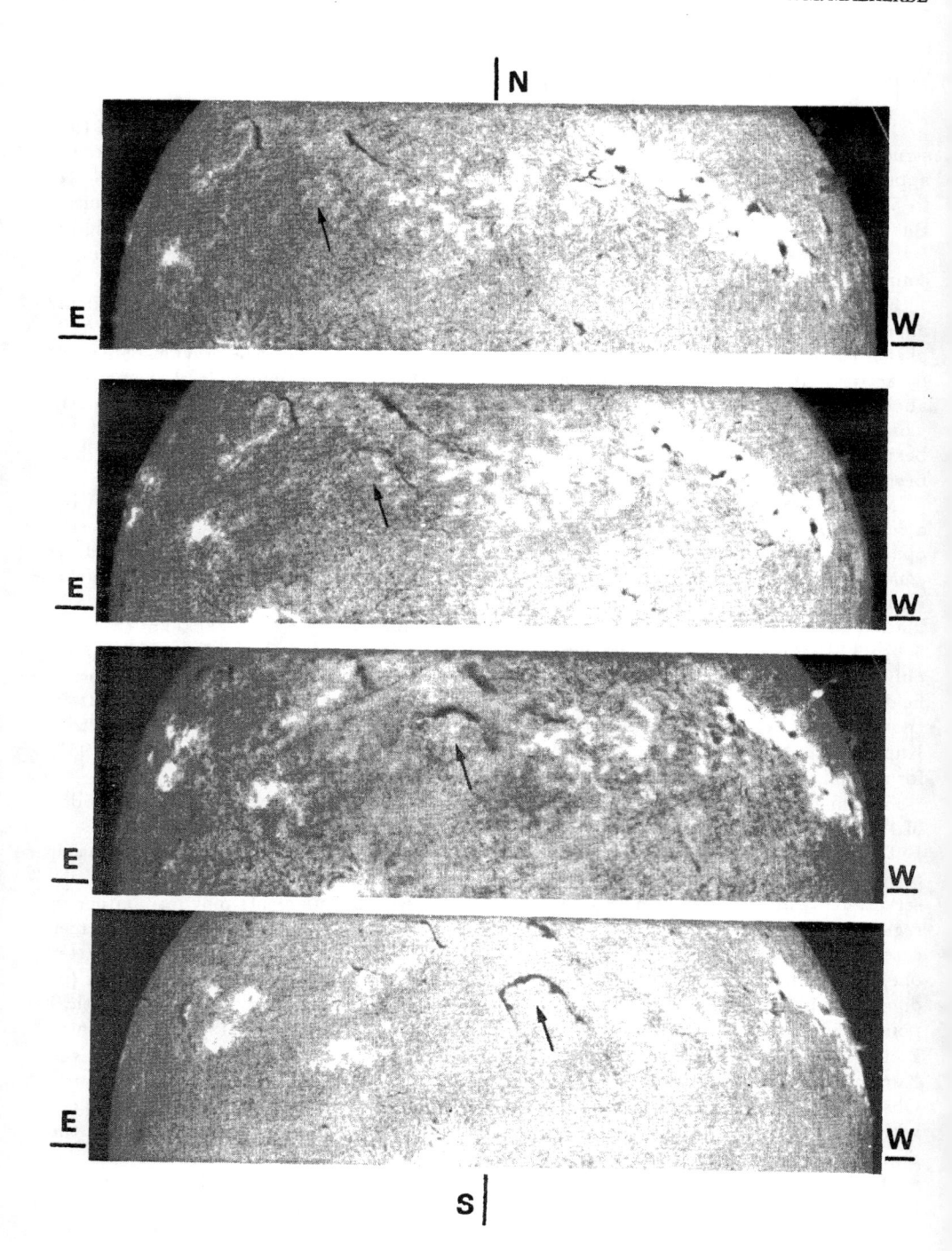

CHAPTER 5: THE FORMATION OF SOLAR PROMINENCES

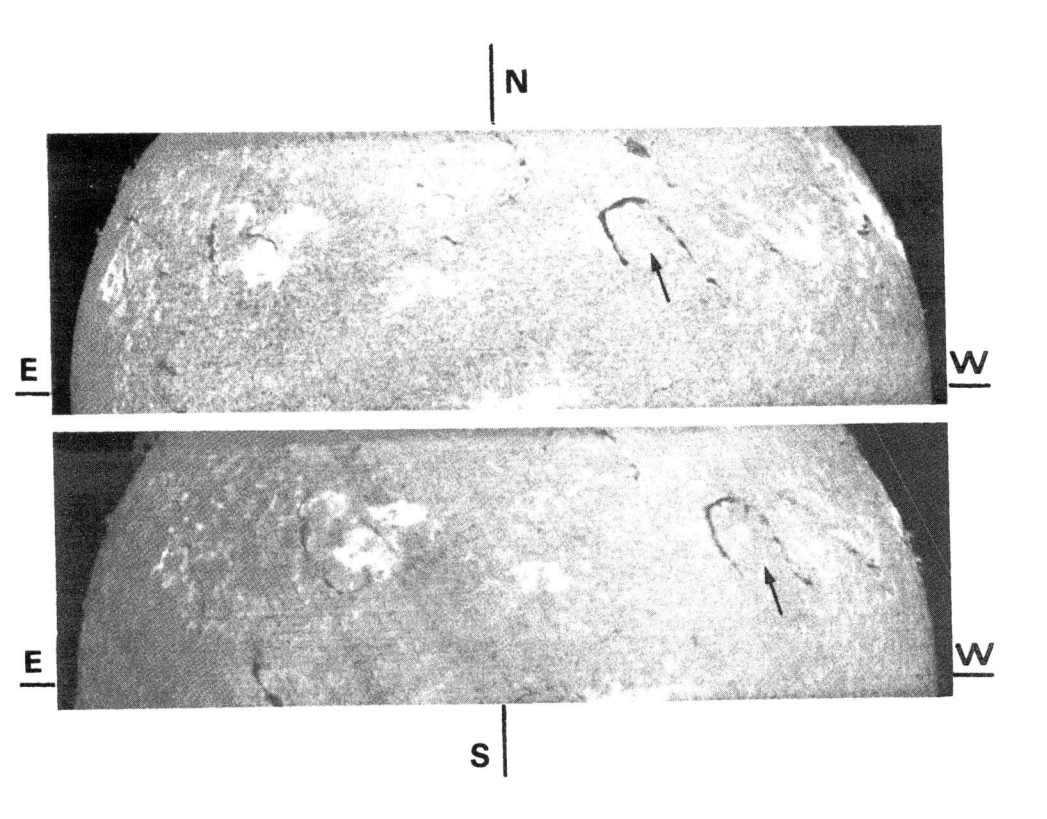

Figure 5.1: Formation of a quiescent filament (April 10, 11, 12, 13, 14 and 15, 1980) seen in the centre of $H\alpha$ line (6563 Angströms). The arrows indicate the location of the appearing filament. This typical event takes several days, or more. The different parts of the filament do not appear together at the same time: first the western part becomes visible; then the filament extends and the eastern part forms. Clearly the formation of the feet occurs in $H\alpha$ before the appearance of the filament itself, as almost regularly spaced condensations of dark material above the photosphere. The arches of cold material between feet seem to form later (from a Meudon spectro-heliograph).

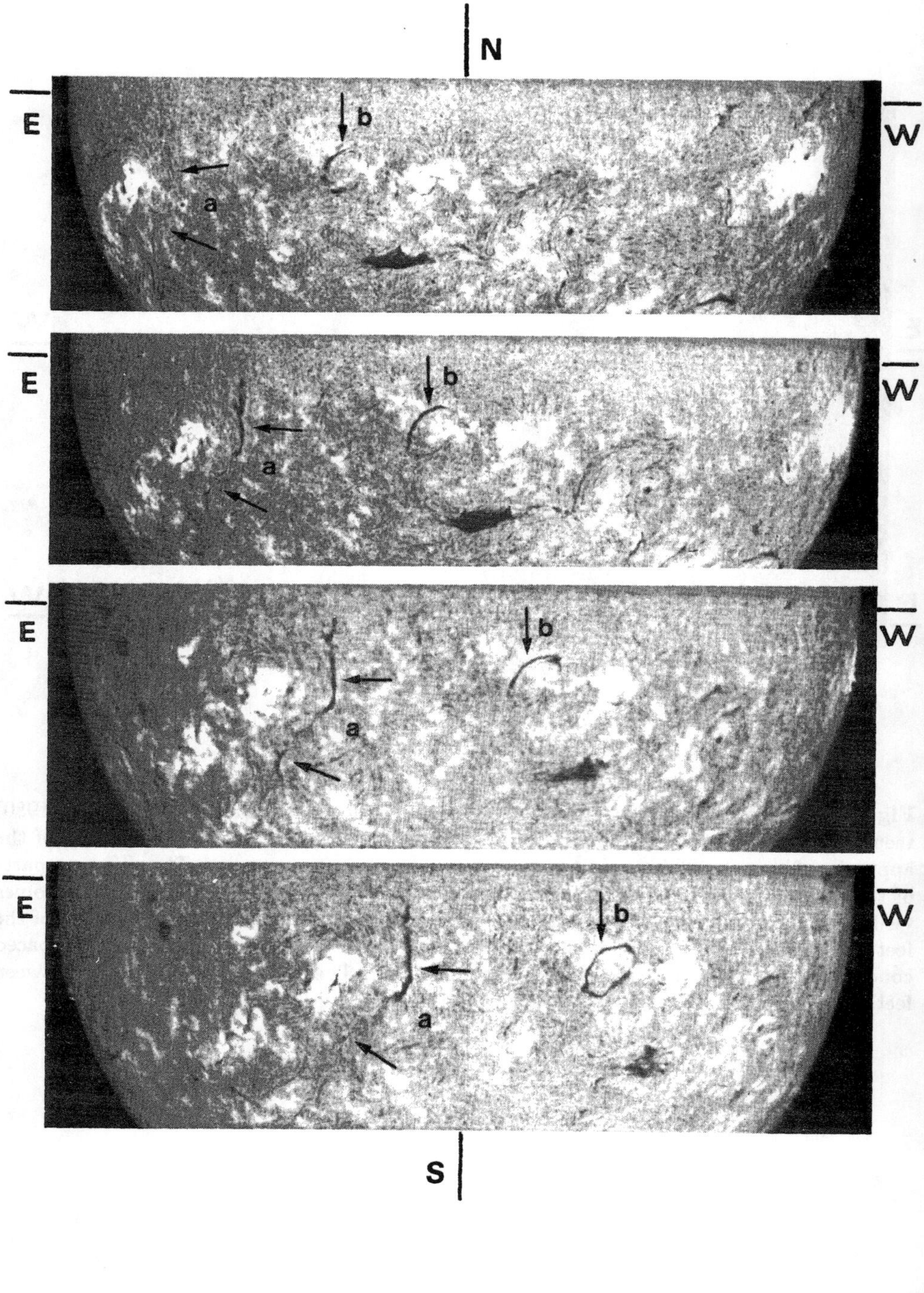
N
E
W
a
b
S

CHAPTER 5: THE FORMATION OF SOLAR PROMINENCES   121

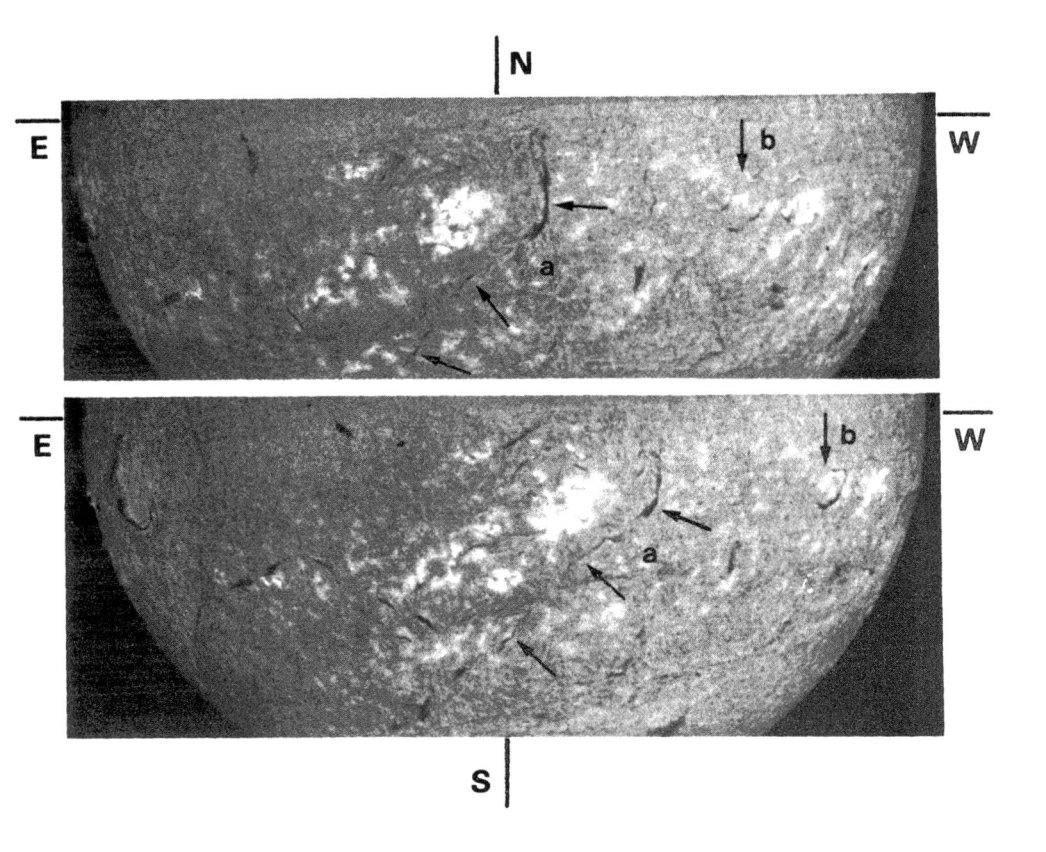

Figure 5.2: Formation of a quiescent filament (a) and winking of a small plage filament (b) (June 4, 5, 6, 7, 8 and 9, 1980) seen in the centre of the $H\alpha$ line. The arrows indicate the location of the appearing quiescent structure or the winking one.
The formation of the quiescent filament (a) takes several days. On June 4, a remarkable orientation of fibrils (which are small threads or low-lying loops of dark material generally parallel to the direction of the magnetic field) is clearly visible before the appearance of the filament: this is the signature of the filament channel, which is a region located at the boundary between two opposite magnetic polarities. The orientation of fibrils along such a channel often precedes the formation of a quiescent filament. On June 9, the filament has considerably extended towards the South and the location of the feet is well visible.
Filament (b) is a winking plage filament: alternatively heating and cooling, it appears and disappears in $H\alpha$ due to nearby activity. This is a very common phenomenon in plage filaments which are close to active centres; it involves a time-scale between one hour and one day (from a Meudon spectro-heliograph).

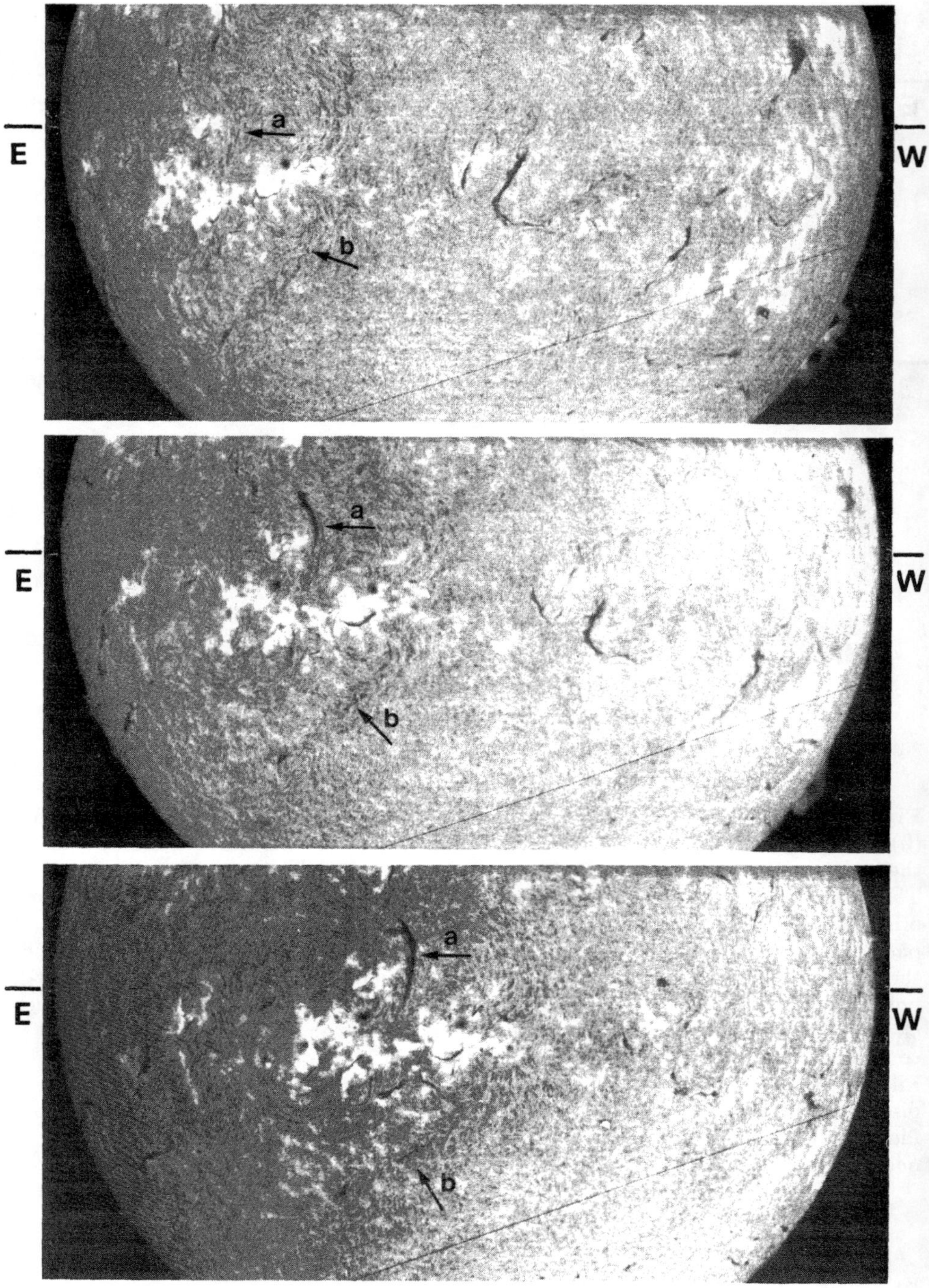
a
E
W
b
a
E
W
b
a
E
W
b

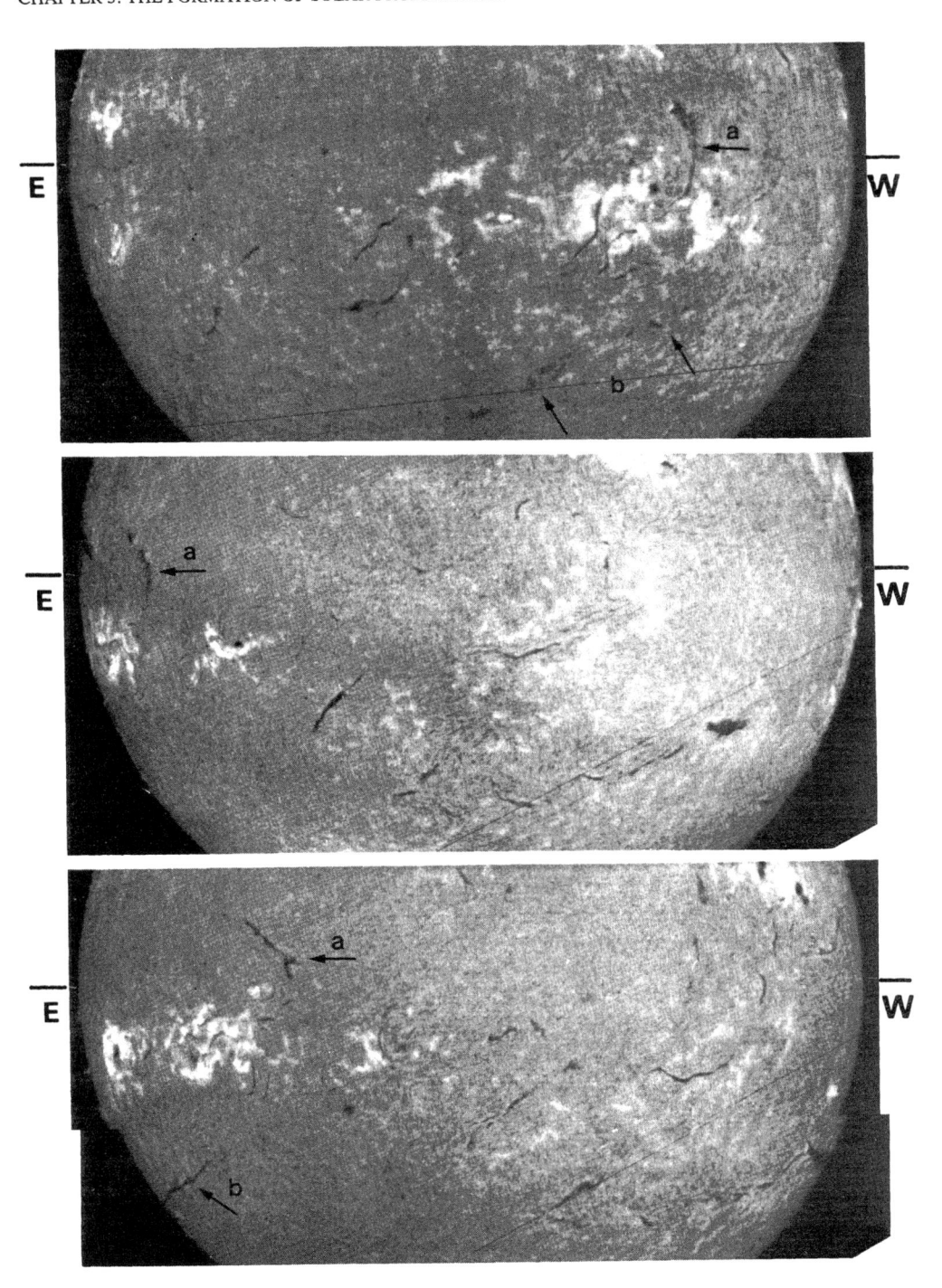

CHAPTER 5: THE FORMATION OF SOLAR PROMINENCES 123

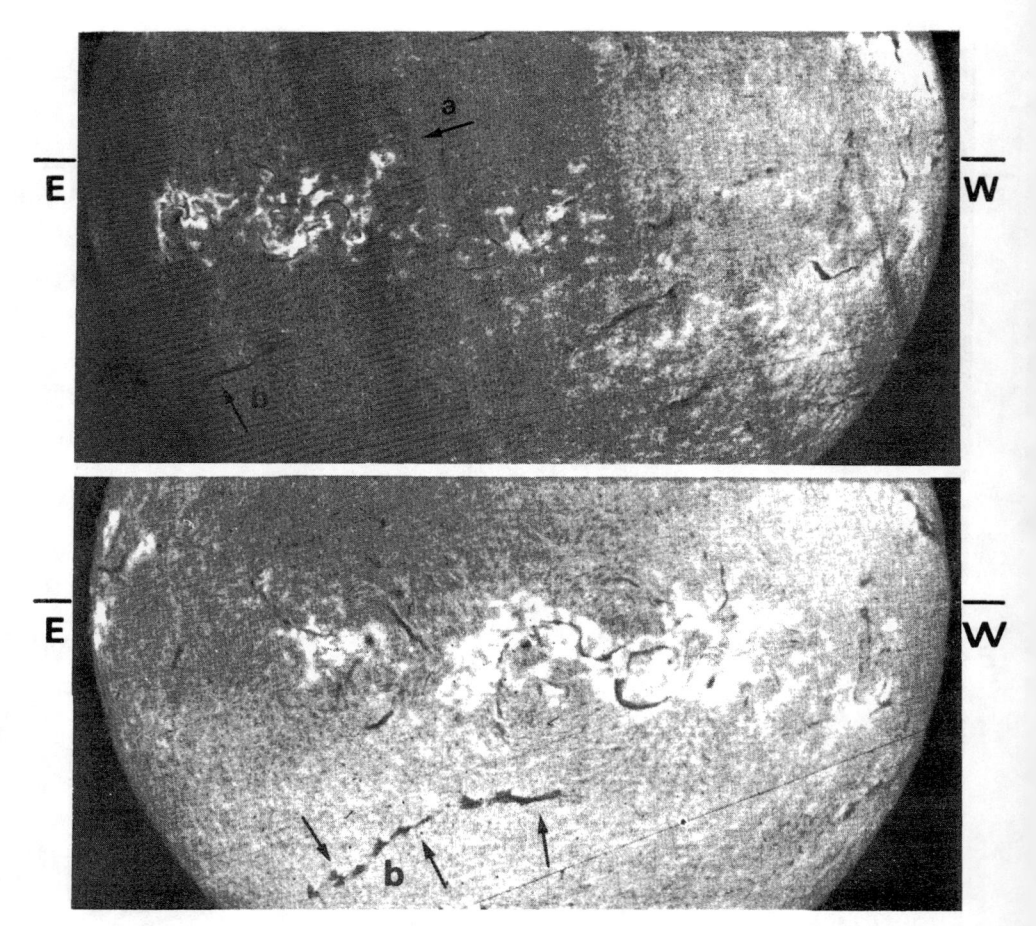

Figure 5.3: "Apparition Brusque" (AP) of a plage filament (a) and formation of a quiescent filament (b) (January 30 and 31, 1981; February 1, 4, 23, 25 and 26, 1981; March 2, 1981) seen in the centre of the $H\alpha$ line. The arrows indicate the locations of the filaments. On January 30, a peculiar orientation of fibrils (filament channel) precedes the appearance of filament (a). This is an "Apparition Brusque", in contrast to the well-known "Disparition Brusque" (DB) phenomenon. The AP physically differs from the formation process of a new filament: it is in fact the reappearance (and the cooling) of an existing filament which was previously heated due to activity in the neighbourhood and so was no more visible in chromospheric lines. It must be noticed that the different parts of the structure reappear together at the same time (this is not the case in filament formation). The time-scale for the cooling is typically between one hour and one day. On February 26, a new DB has occurred (but the filament is probably still there and visible in UV lines): recurrent AP and DB's are probably a good explanation for the winking of plage filamants.
Filament (b) is a newly forming one; here the duration of the process is several weeks. The filament channel exists on January 30; on February 4, only feet are visible; and on March 2, the filament body has formed in $H\alpha$ (from a Meudon spectro-heliograph).

CHAPTER 5: THE FORMATION OF SOLAR PROMINENCES   125

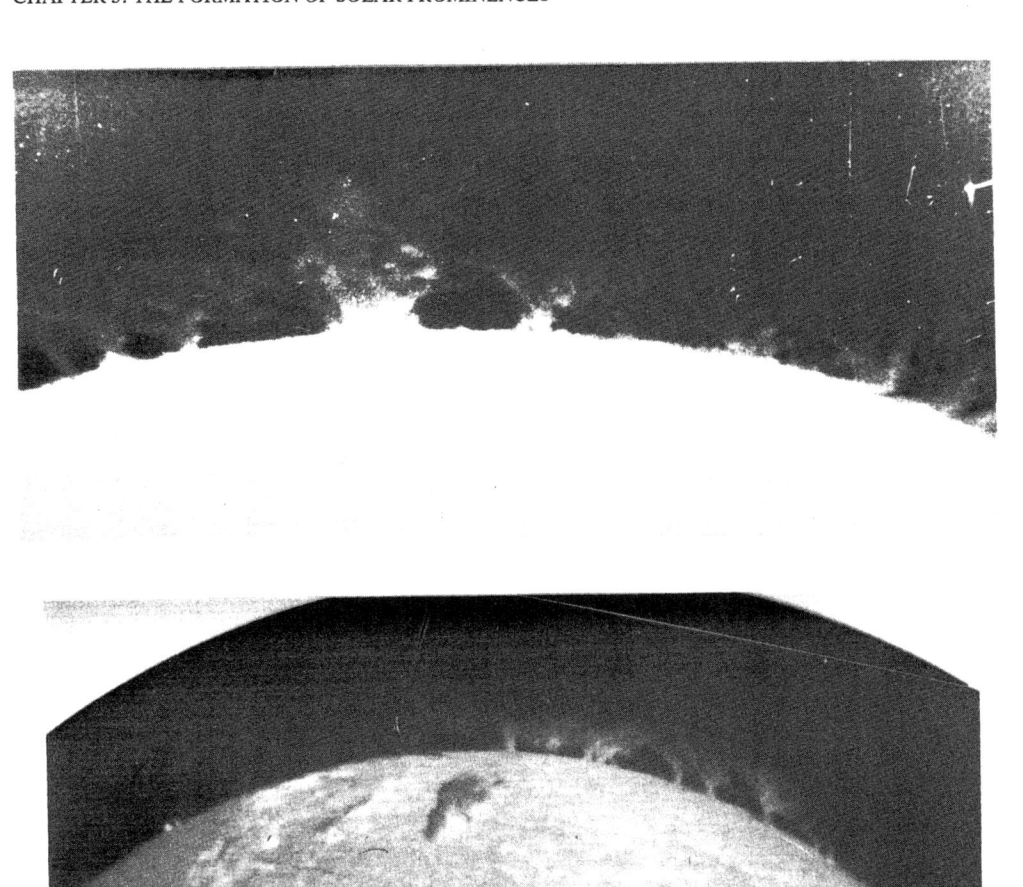

Figure 5.4: Quiescent prominences seen in the centre of the $H\alpha$ line (top: February 5, 1959; bottom: August 8, 1979). Once the prominence has formed, almost regularly spaced feet with arches and loops of dense material between them are visible. But the stability of loops 100 times higher than the pressure scale-height is still a puzzle; if the magnetic field is parallel to the dense loops, the structure cannot be static, but must be highly dynamic. Unfortunately, the magnetic topology and the velocity field of individual loops in prominences is not yet known: it is the key to the physics of prominences (thermal and mechanical equilibrium). The feet seem to be extremely important in the stability of the structure: when the distance between feet differs too much from the height of the prominence, it cannot be stable for a long time (from a Meudon spectro-heliograph).

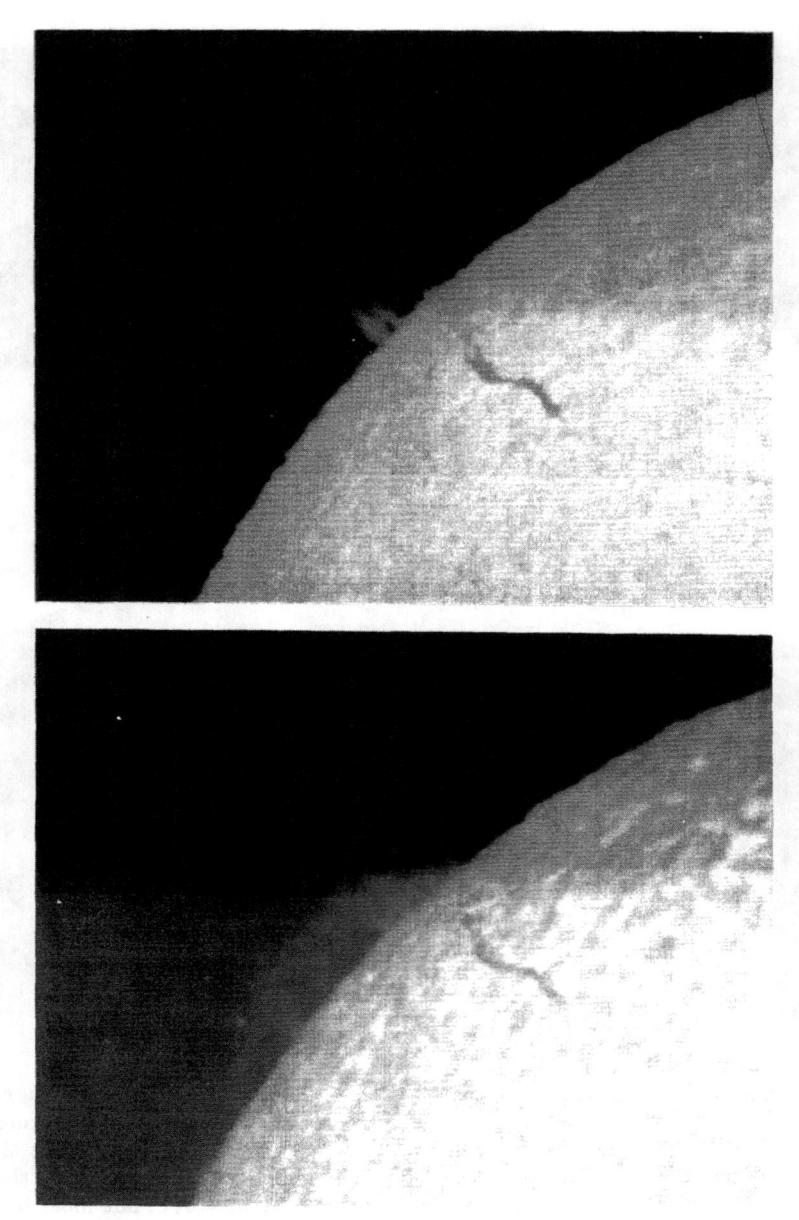

Figure 5.5: A quiescent prominence (June 5, 1982) seen in the centre of $H\alpha$ (top) and Ca II K_3 (3934 Angströms, bottom) lines. While the $H\alpha$ picture exhibits mainly vertical threads, the K_3 line rather shows dense loops; this means that prominence material is not homogeneous in temperature. At the moment, the thermal structure of individual loops is not understood at all: a prominence may consist of isothermal loops at different temperatures, or non isothermal loops (from a Meudon spectro-heliograph).

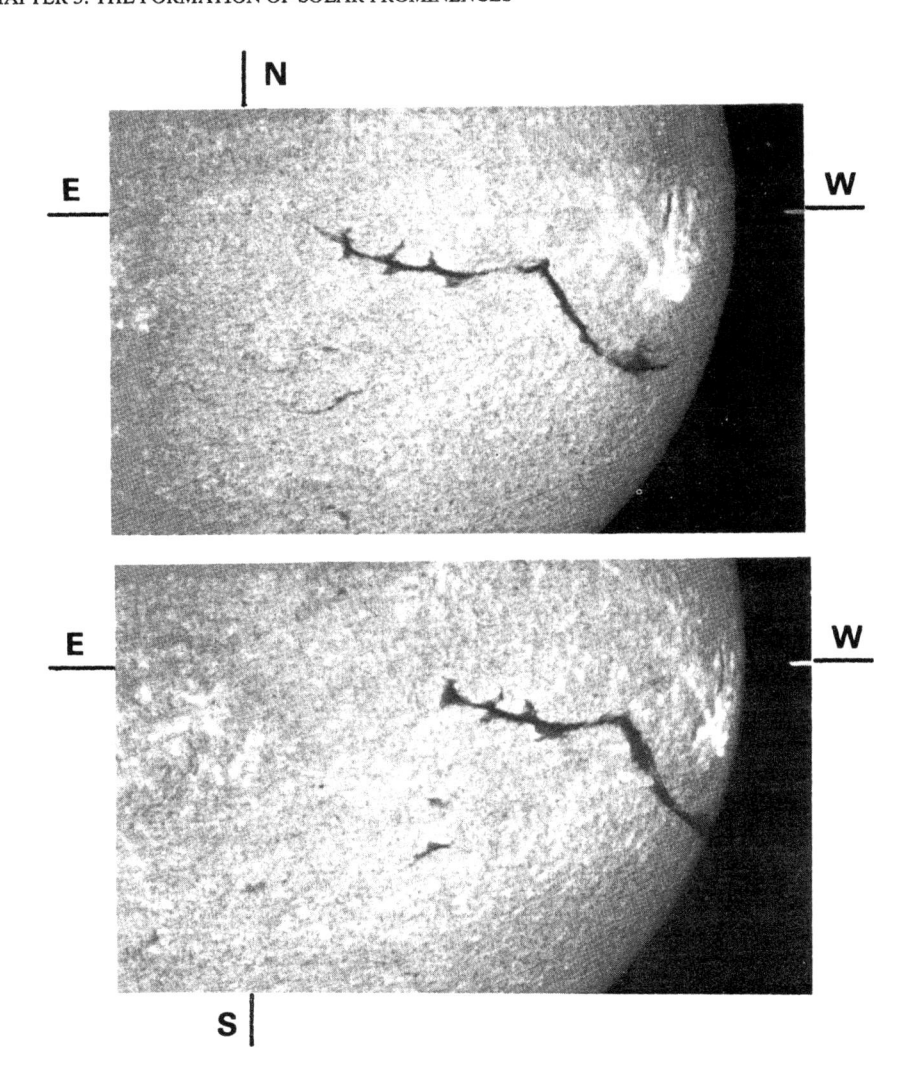

CHAPTER 5: THE FORMATION OF SOLAR PROMINENCES 127

Figure 5.6: A quiescent filament seen in the centre of the $H\alpha$ line (May 3 and 4, 1973). The feet are exceptionnally well visible; they are anchored in the photosphere on both sides of a magnetic inversion line. It is not known, at the moment, if the feet are located at the boundary of supergranules or inside. The measurements made with magnetographs show that the magnetic field at the feet is very weak and almost undetectable (smaller than 10 Gauss). As the filling factor is unknown, it is very difficult to have a good idea of the magnitude of the field (from a Meudon spectro-heliograph).

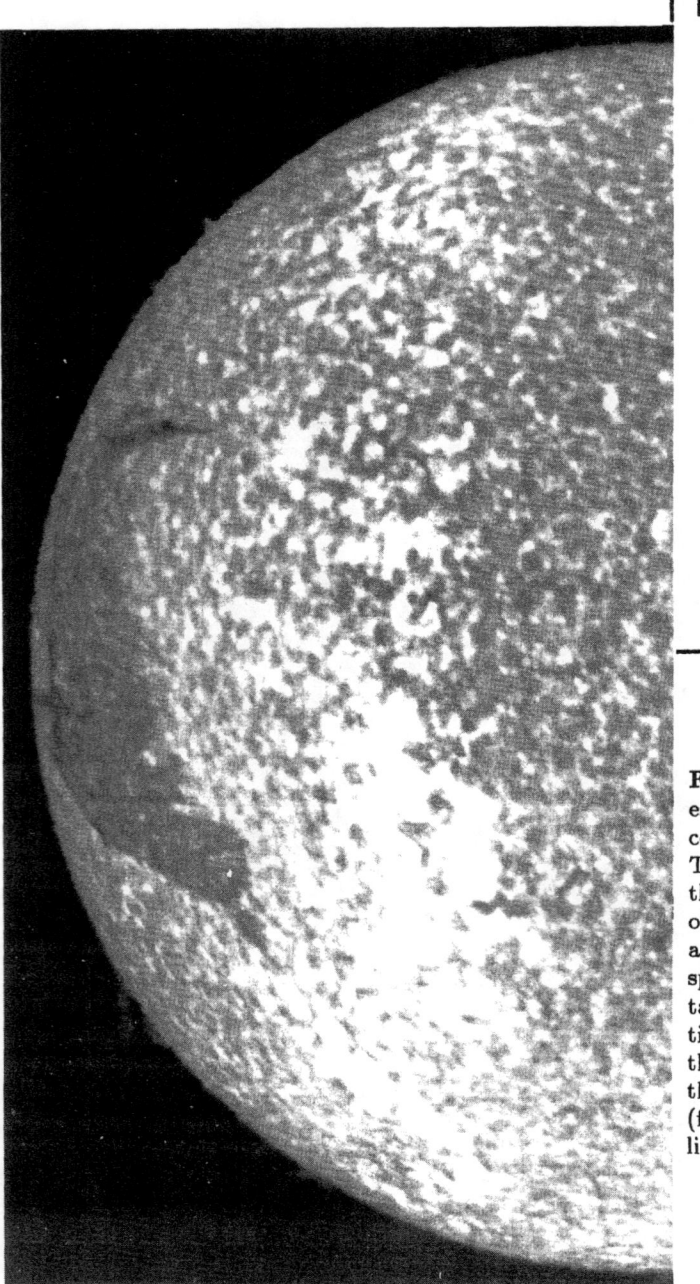

Figure 5.7: A huge quiescent filament seen in the centre of the Ca II K_3 line. This picture shows clearly that a filament is made up of a set of cold loops which are anchored in the photosphere. It is a very difficult task to find out the location of the foot-points of the loops with respect to the supergranular network (from a Meudon spectro-heliograph).

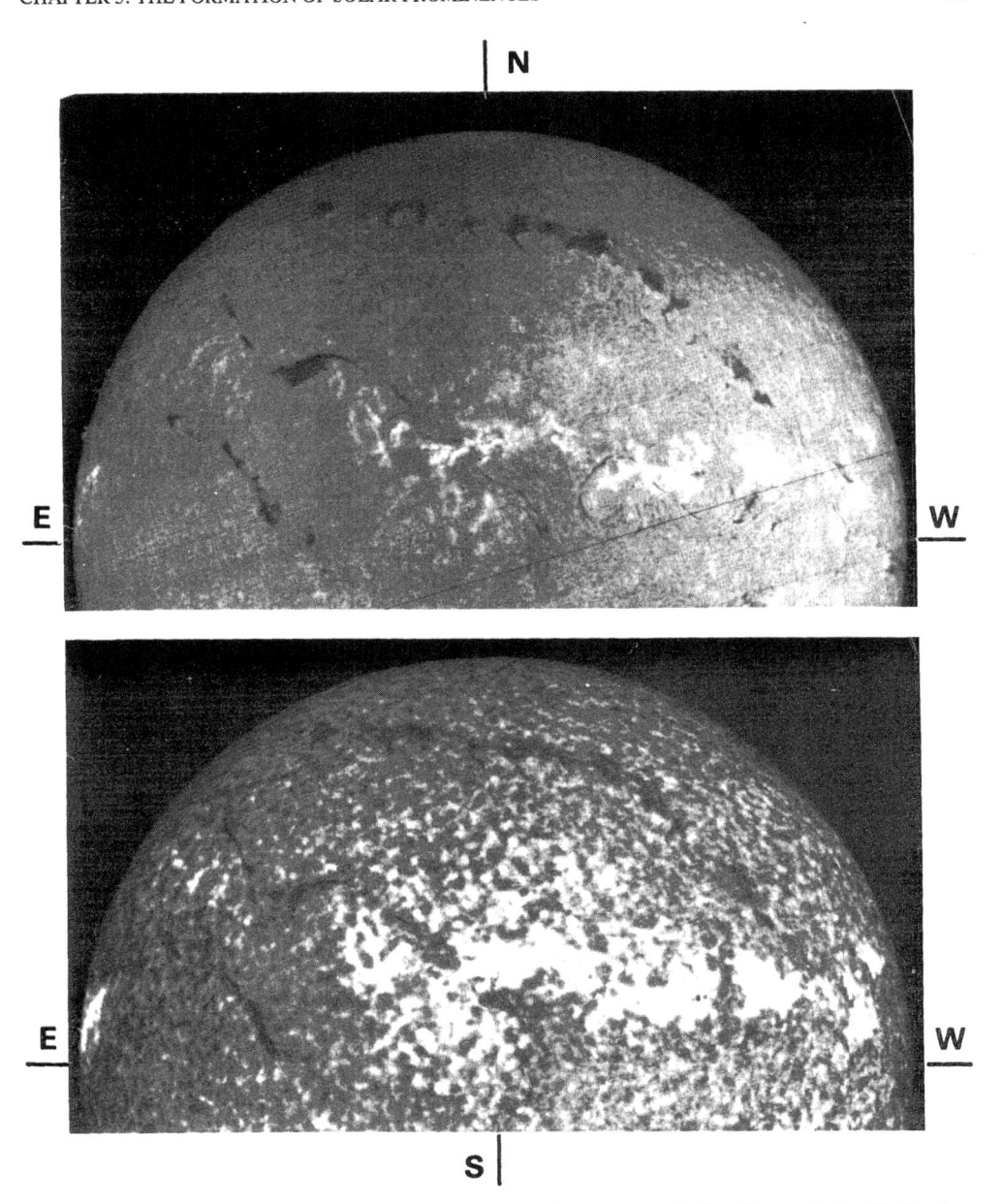

CHAPTER 5: THE FORMATION OF SOLAR PROMINENCES 129

Figure 5.8: A large quiescent filament seen in the centre of the $H\alpha$ and Ca II K_3 lines (May 26, 1969). On the $H\alpha$ picture, only the feet are visible. But, in K_3, both feet and filament body are present; the structure seems to be more homogeneous in K_3 than in $H\alpha$. This example suggests that the feet and dense arches between them form simultaneously, even if they do not appear in $H\alpha$ at the same time, probably due to temperature gradients. This is an important point for the models (from a Meudon spectro-heliograph).

5.3 Description of Main MHD Instabilities Involved in Prominence Formation

5.3.1 RADIATIVE THERMAL INSTABILITY

The basic MHD instability involved in plasma condensation is the radiative thermal instability (Field, 1965; Priest, 1982). It is the consequence of the shape of the cooling function $Q(T)$ (see Priest, 1982, or Hildner, 1974) which has a maximum at transition zone temperatures ($10^5 K$). It is possible to show, using orders of magnitude derived from energy equation 1.4, that there exists roughly a cold thermal equilibrium (at prominence temperatures) and a hot one (at coronal temperatures). The equilibrium temperatures are given, at constant gas pressure, if one neglects conduction, by:

$$\frac{Q(T)}{T} = \frac{k_B h}{Pm}$$

where T is the temperature, P the gas pressure, m the mean atomic mass and k_B Boltzman's constant. h represents the heating rate per unit mass; it is generally assumed constant and may be, for simplicity, computed by reference to coronal values so that $h = (P_c m/k_B)(Q(T_c)/T_c)$, subscript c referring to the corona.

Let us now consider a simple 1D model of a non-twisted magnetic loop of length L at constant gas pressure, with boundary conditions at the foot-points given by $T = T_c$ (T_c being the coronal temperature), and let us study the evolution of the temperature of the loop summit under an isobaric perturbation. The energy equation, in order of magnitude, reduces to:

$$\rho C_p \frac{dT}{dt} = h\rho - \rho^2 Q(T) + k_0 T^{5/2} \frac{T_c - T}{L^2}$$

where ρ is the mass density related to the temperature by $\rho T = $ constant, C_p is the specific heat at constant pressure ($C_p = (k_B/m)(\gamma/\gamma - 1)$ with $\gamma = 5/3$), and $k_0 = 3\ 10^{-11}$ $Jm^{-2}s^{-1}$ is the thermal conductivity constant (t is the time).

One may show that the cold equilibrium is stable, but the hot one is unstable without sufficiently strong conduction (as the plasma cools, it radiates more, because the radiative losses are maximum at transition zone temperatures). Conduction may stabilize radiation (this is the case in the corona), but the stabilizing effect decreases as the length (or the twist or the shear) of magnetic lines increases, or as wave heating (generally assumed constant per unit mass) decreases. The simple model above shows that thermal instability is triggered when the ratio of radiative to conductive time-scales becomes smaller than unity.

The linear radiative time-scale is

$$\tau_R = \frac{P_c}{\rho_c^2 Q(T_c)}$$

while the linear conductive time-scale is

$$\tau_C = \frac{P_c L^2}{k_0 T_c^{7/2}}$$

CHAPTER 5: THE FORMATION OF SOLAR PROMINENCES   131

where P_c, ρ_c, T_c are, respectively, the coronal pressure, density and temperature, and L is the length of field lines. The linear radiative time-scale is, using classical coronal values, of the order of a few hours, but cooling may evolve much faster in the nonlinear phase (factor 3 or more). Condensation may occur approximately when $\tau_R/\tau_C < 1$, or when the length of the loop L exceeds a critical value L_m given by (in meters):

$$L_m = 3.21 \ 10^{-14} T_c^{13/4}/P_c$$

The critical half-length is about 50000 km for a loop or an arcade with typical coronal conditions ($T_c = 10^6$ K and $P_c = 0.01$ Pascals).

5.3.2 RESISTIVE INSTABILITIES

Resistive instabilities become important on singular layers where $\nabla \times (v \times B)$ vanishes, v and B being, respectively, the velocity and magnetic field. This is, for instance, the case for current sheets, which are good candidates for the formation of condensations (a summary of current sheet behaviour is presented by Priest, 1982). We now need to introduce the diffusive linear-time scale:

$$\tau_D = \frac{W^2}{\eta}$$

and the ideal MHD Alfvén time-scale:

$$\tau_A = \frac{W}{V_A}$$

where W is the width of the current sheet, η the magnetic diffusivity ($\eta = 1/(\mu_0 \sigma)$, $\mu_0 = 4\pi 10^{-7}$ being the magnetic permeability and σ being the electrical conductivity which normally in the corona varies as $8 \ 10^{-4} T^{3/2}$ Mho/m), and V_A the Alfvén speed in the corona ($V_A = B/(\mu_0 \rho)^{1/2}$, and typically 1000 km/s if one takes a magnetic field of 10 Gauss and a density n of 10^9 particles/cm^3). The Alfvén time scale represents the time needed by an Alfvén wave to travel through the sheet. It may be of the order of a few seconds if one takes a sheet width of 5000 km (typical thickness of prominences).

It is also useful to introduce the magnetic Reynolds number R_m as:

$$R_m = \tau_D/\tau_A = \frac{WV}{\eta}$$

where V is a typical velocity. This quantity is called the Lundquist number when $V = V_A$. It is usually about 10^8 - 10^{12} in the corona (but unknown, although probably large, in prominences as the electrical conductivity is not well determined) and shows that pure diffusion is an extremely slow phenomenon in the corona. Furth et al (1963) have suggested that the linear tearing mode instability may produce in current sheets much shorter time-scales (but still long) τ_T of the order of $\tau_A \times R_m^{1/2}$ (about $10^4 - 10^6 \times \tau_A$, numerically 1 day to 1 month) intermediate between the ideal MHD Alfvén time τ_A and the diffusive time $\tau_D = \tau_A \times R_m$. It must be noticed that this time-scale τ_T may be strongly reduced in the presence of anomalous resistivity (due to turbulent motions) or if the current sheet is very thin (a few kilometers). This instability develops long magnetic islands with a wavelength

$\lambda = W R_m^{1/4}$. It converts magnetic energy into heat and kinetic energy much faster than pure diffusion.

The nonlinear evolution of the tearing instability must be studied by numerical means. When two opposite magnetic fluxes approach together (current sheets) at a speed V, one may reach a fast steady-state reconnection regime; this regime is characterized by the value of V. If

$$V_A/R_m^{1/2} \leq V \leq V_{max}$$

V_A and R_m being the Alfvén speed and the Lundquist number at large distances from the sheet, we have the Petschek (1964) regime (with $V_{max} = \pi V_A/(8 \log R_m)$). Typically, V_{max} lies in the range 0.01 V_A and 0.1 V_A, and the magnetic energy is released by the presence of two slow MHD shocks at a much faster rate than the tearing mode time-scale (the Petschek time-scale, normalized to the Alfvén time, is $(8/\pi) \log R_m$ instead of $R_m^{1/2}$, so numerically 50 instead of 10^4 - 10^6).

But, when at large distances from the sheet one has $V > V_{max}$, a steady state is no longer possible: as the reconnection rate is too slow, the magnetic flux accumulates on both sides of the sheet which begins to expand in order to adjust to the new boundary conditions: this is the flux pile-up regime. And when the length L of the sheet becomes too large compared to the width W, the current sheet becomes unstable and produces by reconnection several magnetic islands on a time-scale τ_T. When several islands have formed, they may become unstable and annihilate by the coalescence instability which produces an impulsive bursty regime (Priest, 1984) and releases sporadically a large amount of magnetic energy in a very short time-scale of the order of τ_A. Typically, the coalescence instability occurs when $L/W > R_m^{1/2}$.

5.4 Introduction to Steady Magnetic Reconnection in Current Sheets

5.4.1 INCOMPRESSIBLE AND COMPRESSIBLE THEORIES

Magnetic reconnection (e.g. Priest, 1982, 1984, 1986) is an important phenomenon in solar physics. There exist on the sun two types of reconnection, either forced or spontaneous, according to boundary conditions. Magnetic reconnection may play a major role in heating the upper atmosphere and in solar flares; in prominences, it may be important in inverse polarity structure, since observers suggest steady flows permanently crossing the structure.

The first model for steady-state and incompressible magnetic reconnection is due to Sweet (1958) and Parker (1963), but it exhibits very slow reconnection rates (the reconnection rate represents the speed V, measured at large distances, at which magnetic fluxes approach the current sheet; it is normalized to give the Alfvén Mach number $M_A = V/V_A$). This number, in the Sweet-Parker regime, is $R_m^{-1/2}$, and numerically very small (10^{-4} – 10^{-6}). In the Petschek regime (1964), the reconnection region expands to give two pairs of slow shocks which convert efficiently magnetic energy into kinetic energy and heat. The central diffusive region is Sweet-Parker like, but much smaller and extended by the slow MHD shocks. The maximum reconnection rate is then faster ($\pi/(8 \log R_m)$, and numerically between 0.01 and 0.1). Sonnerup (1970) has proposed a model which allows a high reconnection rate (about 1). The direction of the flow in these two models differs at large distance: it converges towards the sheet (Petschek) or diverges (Sonnerup).

More recently, Soward (1982) and Soward and Priest (1982) put forward models for compressible steady-state magnetic reconnection, in the coplanar or noncoplanar case.

5.4.2 UNIFICATION OF DIFFERENT REGIMES OF INCOMPRESSIBLE MAGNETIC RECONNECTION

Priest and Forbes (1986) have unified the different regimes of incompressible steady-state magnetic reconnection: they have shown that the boundary conditions are extremely important and determine generally the type of magnetic reconnection. The previous solutions (Sweet-Parker, Petschek, Sonnerup, flux pile-up regimes) are in fact particular cases of a more general solution, which can produce a wide set of solutions either with very slow ($M_A \ll 1$) or very fast ($M_A \geq 1$) reconnection rates. Cases with no current (Petschek) may occur in the presence of free-floating boundary conditions; but other cases (with current) could be instead relevant to boundaries with forced flows at large distances from the sheet. The properties of the flow (magnitude and direction, converging or diverging), and the behaviour of the gas pressure and the magnetic field B along the boundaries are essential to discriminate between different regimes of reconnection. One may have compression or expansion regimes when the plasma pressure respectively increases or decreases towards the sheet. Expansions may be fast or slow when the magnetic field decreases or increases towards the sheet. And compressions may be fast or slow when the magnetic field increases or decreases towards it. The Petschek regime separates expansions from compressions.

It must be noticed that resistive instabilities as well as steady-state magnetic reconnection play an important role in the formation and equilibrium of inverse polarity models of prominences (such as the Kuperus-Raadu model). The formation of prominences in current sheets involves resistive instabilities and their nonlinear development. The dynamical equilibrium of inverse polarity filaments involves large-scale photospheric convection (with converging motions at the foot-points of magnetic lines of the order of 10 m/s): it implies that reconnection at the X-line should be forced and controlled by photospheric motions at the feet of field lines. This forced regime may be compatible with the slow steady-state Sweet-Parker model.

But, when one considers two-ribbon flares, generally associated with the eruption of a plage filament, the high speed of the bright $H\alpha$ ribbons moving apart (at a speed greater than 10 km/s) is compatible with the spontaneous fast steady-state regime of magnetic reconnection proposed by Petschek (1964), which is expected to occur above in the corona at an X-type neutral line (Kopp and Pneuman, 1976; Forbes and Malherbe, 1986; Forbes et al, 1988).

5.5 Static Models

For this section, the reader is referred especially to Priest (1982, chapter 11) who presents models of the thermal aspects of prominence formation (these models are called static here, because they do not incorporate the dynamics of the condensation process).

5.5.1 CONDENSATION IN A LOOP

Several authors, such as Hood and Priest (1979), Chiuderi et al (1981), Einaudi et al (1984), She et al (1986), have studied thermal instability in magnetic loops. Hood and Priest (1979) have shown that, when either the external gas pressure P, or the length L of field lines increases (due to motions of foot-points or due to variations of the twist), or the heating decreases, a thermal instability may occur above critical values, and the core of the loop cools. Below these values, one has an equilibrium between conduction and heating; above, one has equilibrium between radiation and heating. Hence, thermal energy is redistributed in this non-linear process, due to the presence of catastrophe or non-equilibrium solutions. An (1984, 1985) looked at the condensation modes in cylindrical plasmas (loops) and examined the effect of the shear (An, 1986). The thermal behaviour of all these models shows in fact that the evolution of boundary conditions is very important in the thermodynamics of the solar plasma.

5.5.2 CONDENSATION IN AN ARCADE

Priest and Smith (1979) have studied thermal instability in 2D magnetic arcades, with coronal-like boundary conditions at the base. They have shown that, when the length L or the shear of field lines increases, or when the external gas pressure is enhanced, or when wave heating h decreases, thermal instability occurs in a finite region of the arcade. Then the instability pumps material on both sides of the magnetic arcade and a siphon flow regime is established. This is an interesting scenario, because observations show that the shear is always present when condensations appear and is a basic property of filaments. Also, this model explains the finite height of prominences (when one considers only the overall structure): above the photosphere, the small length of magnetic lines (strong conduction) prevents the occurrence of the thermal instability. At high altitude, conduction is reduced, but radiative losses $(\rho^2 Q(T))$ are also very small as density decreases with height: the net result is to prevent, here again, thermal instability.

5.5.3 CONDENSATION IN A SHEARED MAGNETIC FIELD

Chiuderi and Van Hoven (1979) and Sparks and Van Hoven (1985) considered thermal instabilities of the coronal plasma in sheared magnetic fields; Van Hoven et al (1984) studied the interaction between radiation and tearing, while Tachi et al (1985) investigated the effects of viscosity in radiative and reconnection instabilities of sheared fields; Van Hoven and Mok (1984), and Van Hoven et al (1986), incorporated the effect of anisotropic thermal conduction and found unstable modes due to the perpendicular component.

Steinolfson (1983) found the interesting result that, at high temperatures ($T > 3\ 10^5 K$), there exist two unstable modes: the radiative mode grows faster than the tearing mode (if the magnetic Reynolds number exceeds about 10^7). But, at low temperatures ($T < 8\ 10^4 K$), there is only one unstable mode, the tearing mode, and the plasma is thermally stable. Steinolfson (1983) and Steinolfson and Van Hoven (1984) found that the radiative tearing mode is hybrid: at classical coronal Reynolds numbers, it exhibits time-scales which are relevant to the ideal radiative time (a few hours), but it produces also a high reconnection rate (30 % of the ideal tearing mode, which is, in principle, much slower than the ideal radiative mode): this may be explained by the decrease of the temperature at the X-type neutral points which enhance the resistivity. In contrast, the temperature increases at O-type neutral lines: this is the opposite of what is found in the non-radiative tearing.

More recently, interactions between radiation and magnetic reconnection in current sheets have been numerically studied by Malherbe (1987) and Forbes et al (1988); the results are summarized in Section 5.7.

5.5.4 CONDENSATION IN A CURRENT SHEET

Kuperus and Tandberg-Hanssen (1967) suggested the possibility to form a quiescent prominence in a vertical current sheet. Such a sheet could form after an instability of a magnetic arcade (such an eruptive instability) opening the field lines. Smith and Priest (1977) have investigated the thermal behaviour of a current sheet (with coronal-like boundary conditions at the ends of the sheet), and found it possible to condense the plasma when the sheet elongates (which reduces thermal conduction from the ends of the sheet). The critical length decreases as the external (coronal) magnetic field B_c increases, since the internal pressure P of the sheet is given by $P = P_c + B_c^2/(2\mu_0)$ (so that when P increases, the cooling of the sheet is enhanced).

5.6 Dynamic Models: Injection of the Chromosphere into Closed Loops

There are two main ways to form a prominence: by evaporation or injection of the chromosphere into the corona (by for example a spicule or surge mechanism), or by condensation of the coronal plasma itself. We summarize in this section recent 1D and 2D results obtained with numerical simulations of ideal MHD equations (conservation laws of mass, momentum and energy in the presence of magnetic fields). First of all let us consider injection mechanisms.

Injection processes can be subdivided into surge-like and evaporation-like models. An et al (1986) suggested that material is ballistically launched from the chromosphere to the corona at the speed of spicules, while Poland and Mariska (1986) showed that a sustained heat release in a loop may give rise to an evaporation from below followed by a thermal instability at the top of the loop.

5.6.1 SURGE-LIKE MODELS

It seems likely that the mass in a huge prominence has its origin in the chromosphere, and that a siphon mechanism is responsible for the transfer of mass from the solar surface to the corona. The first model for this was the well-known steady-state siphon flow model of Pikel'ner (1971), in which plasma evaporated from the chromosphere goes up along closed magnetic tubes and condenses as it approaches the top of a magnetic arcade. Ribes and Unno (1980) and Uchida (1979) proposed also stationnary models belonging to this class. Démoulin and Einaudi (1987) have shown numerically (by a 1D model with a complete energy equation) that, effectively, during the evolution of a thermal instability at the top of a loop, a mass flux from the photosphere to the corona may be induced, under certain conditions on both the photospheric boundary, the length of the loop and the initial summit temperature.

Wu et al (1987) have investigated a chromospheric injection process, but have demonstrated that this mechanism alone cannot account for quiescent prominence formation. They have solved numerically the 2D radiative-conductive MHD equations, starting with an initial magnetic arcade (Figure 5.9), and have shown that the formation of a Kippenhahn-Schlüter filament at the top of magnetic loops is a dynamical process, as observed, which requires both condensation of the top of the loop and chromospheric injection to supply enough material, in order to achieve the formation of a dense structure. They suggest (Figure 5.9) that active region filaments could be formed by asymmetric injection into the arcade (for instance by spicules, which are fast upward jets of chromospheric material located at network boundaries), while quiescent objects could be due to symmetric injection (for instance by a slow evaporation of the chromosphere). But the asymmetric spicule mechanism requires a delicate balance between injection speed, density and overlying magnetic field strength, which implies that plage filaments cannot form above every neutral line and must have a transient nature. The symmetric evaporation process (not described in the model) requires a narrow range of field strengths (a few Gauss) as observed.

An et al (1988) have incorporated into this injection model the effects of shearing the magnetic field lines (since we know that the shear is a basic property of quiescent prominences), and also the effects of draining field lines and material due to converging motions at the chromospheric level (since we know that converging flows are predicted by theorists and strongly suspected by observers to occur below quiescent filaments). These authors have shown that the shearing and converging velocity, as well as the shear magnitude, are important parameters for the understanding of normal polarity (KS) prominence formation. They conclude, for instance, that the shear motions are able to induce in the arcade a region characterized by a high current and mass density, and low temperature, which are necessary conditions to achieve prominence formation.

5.6.2 EVAPORATION MODELS

Poland and Mariska (1986) tried to initiate a prominence in a coronal loop (with a preexisting dip at the top) and found it very difficult without injecting mass into the system using the siphon flow mechanism. They solved the full set of 1D MHD equations, including thermal processes, and investigated the effect of wave heating upon the thermal stability of the plasma, following the idea of Davis and Krieger (1982). As a first step, they slightly reduced the heating in the whole loop (by 1 %) to trigger condensation; since this did not achieve prominence densities, they drove, in a second step, mass upward into the well by heating again only the loop legs (chromospheric evaporation). In this way (by a slight disruption of the heating h in the entire loop followed by a gradual return of energy deposition, but only in the legs) they found it possible to reach densities in the dip as high as $10^{11} cm^{-3}$, with chromospheric evaporation velocities of 2 km/s, in a few hours, which may explain observations of plage or active-region filaments.

But, a second and interesting discovery is that, once the prominence has formed, it is very difficult to destroy, and a large amount of heating is necessary to overcome radiation in the dense plasma. This corroborates recent results obtained by Malherbe and Forbes (1986).

CHAPTER 5: THE FORMATION OF SOLAR PROMINENCES   137

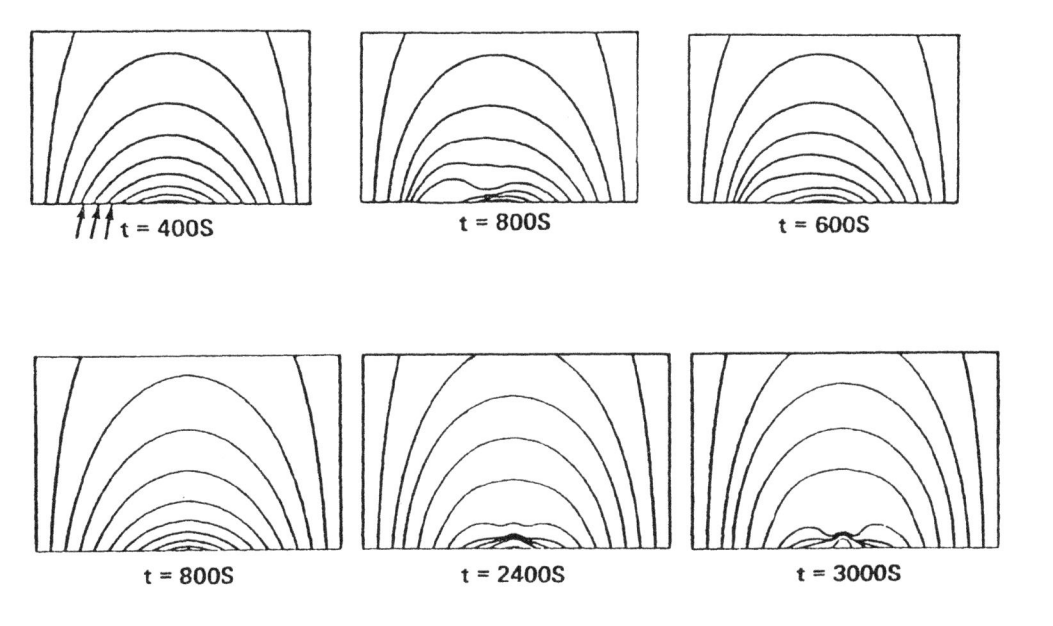

Figure 5.9: A numerical experiment of chromospheric injection into a coronal arcade, showing the evolution of field lines in different cases.

Top: asymmetric mass injection at the speed of spicules (20 km/s) with a density 10 times smaller than the classical coronal one. The figure shows the formation of a nearly symmetrical dip at the top of a field line; then the injected plasma may stably accumulate and condense in the dip to form an active-region prominence.

Bottom: symmetric mass injection of evaporated plasma at the speed of 3.5 km/s and 20 % denser than the ambient medium. It is necessary, to form a quiescent prominence, to form a dip in the field lines in a time-scale shorter than the condensation time-scale, in order to have a stable support (otherwise the dense plasma will flow down under gravity). The field strength must be high enough to support the dense plasma, but not too high to form a dip (here, $\beta = 0.5$). This is why this model predicts a narrow range of supporting field strengths (after An et al, 1986).

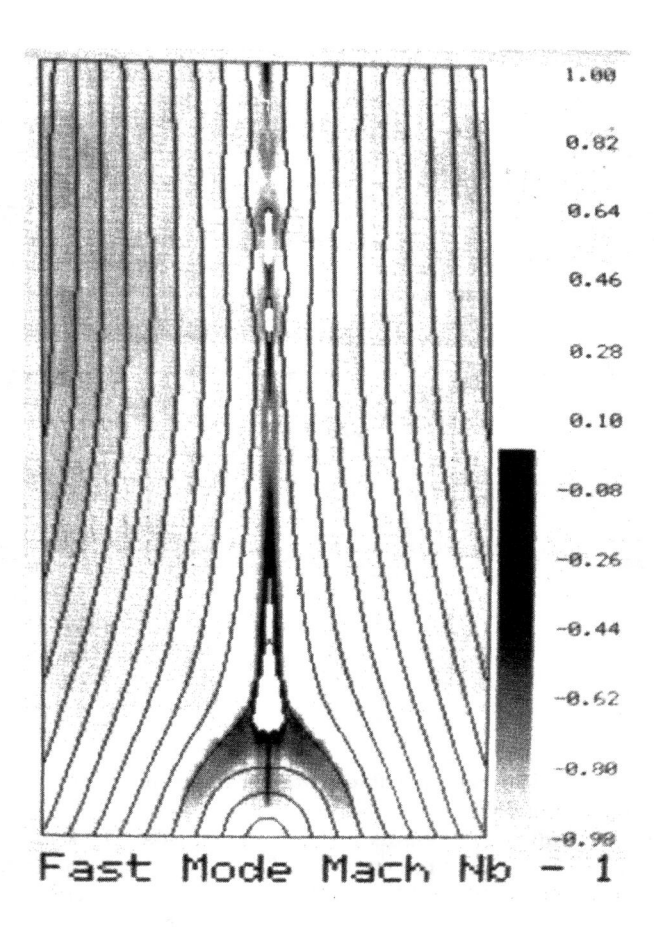

Figure 5.10: A numerical experiment of radiative magnetic reconnection in a vertical current sheet. The magnetic Reynolds number of the box is 800; the width of the sheet is initially 0.075 times the size of the box, and the initial equilibrium is both mechanical and thermal. The plasma β is 0.1. The boundary conditions are free-floating everywhere except at the base (line tying condition). The picture exhibits fields lines, plus, in a grey scale, the fast-mode Mach number M_f at the end of the experiment (200 sheet Alfvén time-scales). Super-magnetosonic regions ($M_f > 1$) are white (triangle). A supermagnetosonic (Mach 2) jet originates from the reconnection site (X line) and forms a fast-mode shock when it encounters the top of closed loops formed by reconnection; just below this shock, a stable and cold condensation, of the normal polarity type (KS), has formed inside the loops. At the beginning of the experiment, inverse polarity (KR) condensations have formed but are unstable: since gravity is not included, they have been ejected through the top of the box by the reconnection flow (after Forbes et al, 1988).

CHAPTER 5: THE FORMATION OF SOLAR PROMINENCES

5.7 Dynamic Models: Condensation in Coronal Current Sheets

5.7.1 NUMERICAL SIMULATIONS

Kuperus and Tandberg-Hanssen (1967) suggested that a prominence may condense in a vertical current sheet formed after an instability opening a magnetic arcade (in the post-flare phase for instance). There is also another possibility: when two large convective cells approach together (for instance when the emergence of new active regions push the old ones towards the poles at the beginning of a new cycle), a current sheet may form at their boundary (which may be a good candidate to explain the existence of polar crown prominences).

Malherbe (1987) and Forbes and Malherbe (1986) have numerically solved the 2D resistive-radiative MHD equations in such line-tied current sheets. It is indeed the only way to investigate properly the interactions and the nonlinear effects between magnetic reconnection (and in particular modifications of the Petchek-like regime) and radiation. They have discovered the possibility to obtain both inverse polarity (Kuperus-Raadu) and normal polarity (Kippenhahn-Schlüter) models (the latter involves a shock condensation mechanism: see below). Their initial condition is a purely 2D line-tied and vertical current sheet, which is in mechanical equilibrium, and either in thermal or isothermal equilibrium. Gravity is not included; there is no thermal conduction, and the electric conductivity is taken independent of the temperature. The computations were performed for magnetic Reynolds numbers in the range 400-1000, and for plasma β varying from 0.1 to 1.0. An adjustment for the radiative loss function was required, in order to produce radiative times close to the tearing time of the sheet. The results have shown that, when the plasma β is small (< 0.2), it is possible to get both normal and inverse polarity condensations. The inverse polarity (KR) prominence is located above the X-type neutral line and is unstable and ejected through the top of the box (due to the reconnection flow); the normal polarity (KS) condensation is stable and forms at the top of closed loops formed by magnetic reconnection below the X-line (Figure 5.10). This result suggests that inverse polarity prominences should form much higher than normal polarity ones, and should have a larger vertical extension (in agreement with observations). Between this X-line and the top of the closed loops, there exists a downwards directed supermagnetosonic jet (at about Mach 2); when this flow encounters the underlying reconnected magnetic arcade, a fast-mode shock (Figure 5.10) produces and triggers a thermal instability of the plasma, which is observed to condense below the shock to form the normal polarity (or KS) feature.

5.7.2 THE ROLE OF SHOCK WAVES IN CONDENSATION PROCESSES

The previous numerical work has shown that, when the plasma β is large (> 0.2), the inverse polarity (or KR) condensation still exists, but the supermagnetosonic flow region no longer exists; hence, the shocked region disappears, and there is no longer a normal polarity (or KS) condensation. This gives evidence of a shock condensation mechanism, which has been analytically studied by Forbes and Malherbe (1986) and Forbes et al (1988). There are in fact two shocks: the slow-mode compressible Petschek-like oblique shocks (decreasing the magnetic field and described by the theory of Soward, 1982, and Soward and Priest, 1982) which originate at the X-line and convert magnetic energy into heat and kinetic energy in two reconnection jets (upward and downward); and the fast- mode perpendicular shock (increasing the field) which does not produce as much heat as the slow shocks, but compresses the plasma when the downward supermagnetosonic flow encounters closed field lines. The authors have shown that the combined effect of the two shocks is to compress the plasma by a factor 5; if one notes that the temperature remains almost uniform around the hot reconnection site (due to strong conduction), the presence of the two MHD shocks reduces the radiative time-scale by the same factor, as the plasma passes through the shocks. Forbes and Malherbe (1986) have also studied analytically the effect of an extra parameter not included in the numerical experiment, the transverse magnetic field. They have shown that, when $\beta > 1$, the flow is always submagnetosonic ($V < (V_A^2 + V_S^2)^{1/2}$, where V, V_A and V_S are respectively the flow, Alfvén and sound speeds) whatever is the third component of the field; but, when $\beta < 1$, the flow may become supermagnetosonic ($V > (V_A^2 + V_S^2)^{1/2}$) if the transverse field does not exceed a critical value. This means that this shock condensation mechanism (for triggering the formation of a normal polarity or KS structure) is more relevant to the formation of post-flare loops and plage filaments (which have a small shear) than to quiescent prominences, in which the transverse field component is the most important. This shock condensation process, which predicts for instance the formation of high-density loop prominences during the gradual phase of large two-ribbon flares, is an extension to the Kopp and Pneuman model (1976). It suggests also that plage filaments could be rather of the normal (or KS) type (since they lie in low β regions and do not exhibit a very large shear), while quiescent prominences would correspond better to the inverse polarity (or KR) model, which may appear, as found in the numerical experiment, at higher β values.

5.8 Unsolved Problems

There exist many unsolved problems about prominence formation. Most of the models are interested only in the overall structure and remain 2D; clearly the observations show that prominences are 3D structures in which feet play a major role; the plasma exhibits also an unresolved fine structure, the physical properties of which are almost unknown.

- The formation and the MHD equilibrium of the feet.
  The observations show that feet play a major role in the formation and the equilibrium of prominences. First, feet connect the structure high in the corona to the solar interior: hence, the behaviour of filaments is related to what happens in the photosphere and the convective zone (differential rotation and convection). The formation of prominences begins probably with the formation of the feet (which contradicts the model of Nakagawa and Malville, 1969); also, once the overall structure has formed, the distance between feet is a criterion for stability: when it differs too much from the height of the prominence, it is not stable (Martres, 1988). Moreover, when anchorage in the photosphere is lost, instabilities are often observed (the "Disparition Brusque" phenomenon). Under differential rotation, the motion of the feet deform the overall structure and modify the shear; and singularities of the differential rotation may also be important in the stability of prominences when one foot or more are located in a singular region (such as "pivot points" found by Mouradian et al, 1987). In that sense, the feet impose on the overall structure severe boundary conditions which must

be incorporated in the models.
At the moment, data (magnetic and kinetic) are urgently required to understand better the physics of the feet. The mechanical equilibrium of material inside feet is a puzzle, since their vertical extension is 100 times the pressure scale-height. Magnetic fields play certainly a fundamental role, but the field topology is unknown: is it orthogonal or parallel to the direction of the loops ? In the latter case, the equilibrium cannot be stable; how dynamic is it ? And what about the thermal structure of such loops ?

- The formation and the MHD equilibrium of the fine structure.
It is now probable that a prominence is not "a body, plus feet, plus a fine structure" but that prominences are a set of thin unresolved dense and cold threads or loops (maybe at various temperatures) originating from one foot and going to another in a region of opposite magnetic polarity. Démoulin et al (1988) studied the thermal stability of periodically spaced strands and predicted a size of only 10 km! There is at the moment no good model to explain the mechanical or thermal equilibrium of such a complicated 3D structure. Hence, it becomes now urgent to obtain precise data about physical parameters of the fine structure: size, life-time, densities, temperatures, velocities and magnetic fields.

5.9 Conclusion

Very important improvements of our knowledge of the physics of prominence formation have been made in this decade, both from an observational and theoretical point of view. First, measurements of velocity fields (below and inside filaments, Chapter 2) and magnetic field vector (Chapter 4) have converged towards an inverse-polarity model (such as the Kuperus-Raadu type) for the overall structure of quiescent filaments; a relationship between their dynamics and photospheric large-scale convection (as giant cells) has been suggested. But large uncertainties remain about densities, and this point is crucial for mass balance between prominences and their coronal environment (Chapter 3).
Modern models show how to form essentially plage and normal-polarity filaments (such as the Kippenhahn-Schlüter type), either by chromospheric injection, evaporation, coronal condensation or shock waves, but many physical effects which have been neglected, for simplicity, really need to be included in the theory. Large vector computers allow us now to realize full numerical experiments and investigate nonlinear phenomena, at reasonably high magnetic Reynolds numbers, such as interaction between reconnection and radiation. But there is, at the moment, no good model to explain the formation of quiescent and inverse-polarity prominences.

Lastly, it becomes now important to incorporate the third dimension in the calculations: observations show indeed that the shear is a basic property of filaments; the legs, which are not understood at all by the theory, play certainly a fundamental role both in the formation and the stability of the structure (disappearances). How do they form, what is their magnetic configuration ? Where are they located with respect to the supergranular network ? 3D observations of MHD parameters of feet seen on the disk and at the limb are urgently required, as well as new models taking into account the 3D aspects of prominences: this a beautiful, but difficult, challenge for the next decade.

CHAPTER 6

STRUCTURE AND EQUILIBRIUM OF PROMINENCES

U. ANZER
Max-Planck-Institut für Physik und Astrophysik
Institut für Astrophysik
D-8046 Garching, F.R.G.

The existing models for magnetohydrostatic equilibrium of quiescent prominences are presented and critically discussed. We also investigate how the energy balance can be achieved in these prominences. The main emphasis is on recent developments in these areas of research.

6.1 Introduction

As has been discussed in the preceding chapters solar prominences are thin elongated sheets which reach high up into the corona. Compared to the surrounding corona they are extremely cool and dense. They are globally stationary and long-lived. There seem to exist two classes of prominence: quiescent and active ones; but at present it is not clear whether there is a continuous transition between these two classes. Prominences always occur along lines where the underlying photospheric magnetic field changes sign. The magnetic field vector usually makes a small angle to the prominence's long axis.

The emphasis of this chapter is on models for quiescent prominences; including those in or near active regions. These models must then explain how the large and heavy structures can exist in the tenuous corona. They also must answer the question why they are so cool compared with the surrounding plasma of the corona (the temperature ratio is more than a factor 100). The first theoretical models appeared in the early fifties. Menzel (1951) and Dungey (1953) presented 2-D magnetohydrostatic isothermal configurations. Their models have the property that the density falls off as $\exp(-z/H)$. In Dungey's models H is the pressure scale height of the cool prominence material. Therefore these models do not describe the prominence geometry in a realistic way. Menzel's models leave H as a free parameter, which can, in principle, be chosen as the prominence height. But they give prominences which are as wide as they are high - in contradiction with the observations. A more detailed model of the overall equilibrium and the internal prominence structure was developed by Kippenhahn and Schlüter (1957). They also discussed the stability of prominences. Since these pioneering days a vast amount of publications has appeared in the literature. In this chapter we shall mainly concentrate on the more recent development of this field; a historical account can be found in the excellent text book on prominences by Tandberg-Hanssen (1974).

6.2 Prominence models

6.2.1 GLOBAL STRUCTURE

The problem of prominence equilibria can be divided into two parts. One is the question of what overall magnetic field configurations are capable of supporting the heavy prominence material against gravity high up in the solar atmosphere. The second is to find the internal equilibrium structure. The reason for being able to split the problem up in such a way is the fact that due to the low prominence temperatures the density scale height is orders of magnitude smaller than the dimensions of the prominence itself ($\sim$100 km compared to $\sim$50000 km). Since they are also very long and fairly uniform along their axis two-dimensional models seem to be reasonable. These global models either use current sheets or line currents to represent the prominence.

6.2.1.1 Two-dimensional equilibria.
6.2.1.1.1 Models with Normal Magnetic Polarity.

Models in which the magnetic field emerges from the solar surface on one side of the prominence, penetrates the prominence sheet and returns to the surface on the other side as shown in Figure 6.1a are usually called Kippenhahn-Schlüter configurations. They show no field reversal and no X-type neutral point underneath the prominence. For the description of the different prominence models we shall now introduce a standard system of coordinates in which the prominence lies in the plane $x = 0$, the plane $z = 0$ is the photosphere and y measures the distance along the prominence. Magneto-hydrostatic equilibrium is achieved if

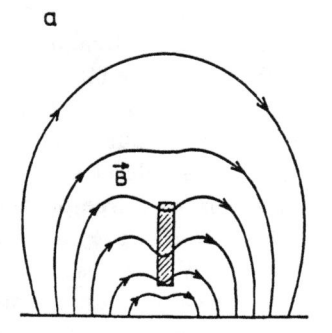

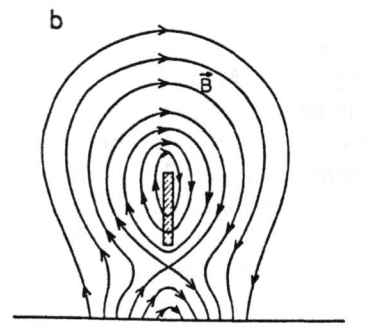

Figure 6.1 Sketch of the possible field topologies: a) Normal Polarity, b) Inverse Polarity.

CHAPTER 6: STRUCTURE AND EQUILIBRIUM OF PROMINENCES

$$\mathbf{j} \times \mathbf{B} = -\rho \mathbf{g} \tag{6.1}$$

holds. In this section we shall deal with models in which the prominence is described by an infinitely thin sheet of matter and electric currents. We define

$$\rho^s := \lim_{d \to 0} \int_{-d/2}^{d/2} \rho \, dx$$

and (6.2)

$$j^s := \lim_{d \to 0} \int_{-d/2}^{d/2} j \, dx.$$

From $\mathbf{j} = \frac{1}{\mu} \nabla \times \mathbf{B}$ and defining $[B_z] = \lim_{d \to 0} (B_z(d/2) - B_z(-d/2))$ one obtains

$$j^s = \frac{1}{\mu}[B_z]. \tag{6.3}$$

Then the equation for vertical equilibrium is the following

$$\frac{1}{\mu}[B_z]B_x = \rho^s g. \tag{6.4}$$

(here $\mu = 40\pi$ for MKS units with B in Gauss). We assume symmetry with respect to $x = 0$ and invariance in the y-direction. Then the prominence is automatically in horizontal equilibrium. At present we also take $B_y = 0$ and outside the prominence $\nabla \times \mathbf{B} = 0$. Models of this kind were first developed by Kippenhahn and Schlüter (1957). They constructed special current sheets starting from potential fields which were symmetric about $x = 0$. The field was then shifted in the x-direction and its mirror image taken on the other side of the prominence (see Figure 6.2). This then gives rise to a sheet current at $x = 0$. One such field is the following: for $x < 0$ the field is produced by a line of dipoles at $z = 0, x = -1$ and a line of fictitious dipoles at $z = 0, x = \alpha$; and for $x > 0$, a line of dipoles at $z = 0, x = 1$ and $z = 0, x = -\alpha$. Along the prominence the field then is given by

$$B_x = \frac{1}{1+z^2} + \frac{\alpha}{\alpha^2 + z^2} \tag{6.5}$$

and

$$B_z = \pm \left(\frac{1}{1+z^2} - \frac{1}{\alpha^2 + z^2} \right) \tag{6.6}$$

with an equilibrium mass distribution

$$\rho^s = \frac{2z}{\mu g} \left(\frac{1}{1+z^2} + \frac{\alpha}{\alpha^2 + z^2} \right) \left(\frac{1}{\alpha^2 + z^2} - \frac{1}{1+z^2} \right). \tag{6.7}$$

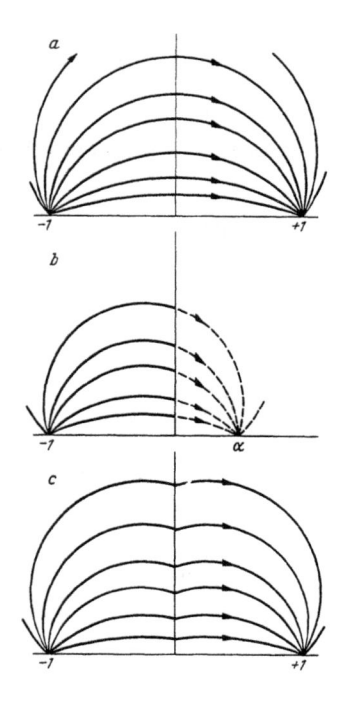

Figure 6.2 Construction of current sheets by mirror-imaging at the plane $x = 0$.

One also has $\int_0^\infty \rho^s dz := M_{tot} < \infty$, therefore such a model has a finite mass (per unit length in y-direction). One can use all kinds of two-dimensional symmetric potential fields for such a procedure and then one has to check afterwards that the resulting field geometries and mass distributions are reasonable.

Two-dimensional configurations can also be obtained with complex-variable functions. With $\xi = x + iz$ any harmonic function $\phi(\xi)$ leads to a potential field if one takes

$$B_x = Im(\phi) \quad \text{and} \quad B_z = Re(\phi). \tag{6.8}$$

Current sheets can be described by functions which have a cut along the z-axis. One can take for example

$$\psi(\xi) = f(\xi) \cdot \sqrt{(p^2 + \xi^2)(q^2 + \xi^2)} \tag{6.9}$$

which leads to a sheet current extending from $z = p$ to $z = q$. Malherbe and Priest (1983) made extended use of this technique. Taking

$$\phi(\xi) = -\frac{B_1}{\xi} \tag{6.10}$$

and

$$\psi(\xi) = iB_0 \frac{\sqrt{p^2 + \xi^2}}{\xi(\xi + ih)} \tag{6.11}$$

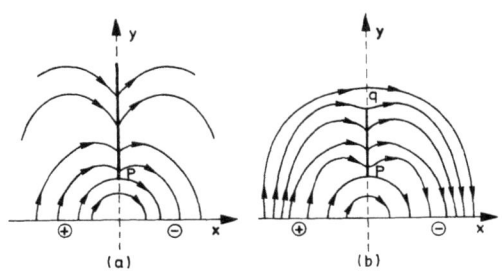

Figure 6.3 Field configuration with current sheets obtained with the help of complex functions: a) current sheet extending from height p to infinity, b) finite current sheet between p and q.

or

$$\psi(\xi) = -B_0 \frac{\sqrt{(p^2 + \xi^2)(q^2 + \xi^2)}}{\xi(\xi + ih)^2} \quad (6.12)$$

they produced the field configurations which are sketched in Figures 6.3a and 6.3b, respectively. Configuration (6.11) is a special case of (6.12) because the current sheet extends to infinity.

The two methods described above both have the shortcoming that one cannot start from observed field distributions to construct the global field. Instead one has to try different analytic functions and then verify that they agree with the observations. Nevertheless they may represent a first step. A method which avoids this problem was developed by Anzer (1972). If the field is symmetric about $x = 0$ and if the prominence extends from $z = 0$ to $z = h$ one can set up the following mixed boundary-value problem: $B_z(x,0) = f_1(x)$ (at the photosphere), $B_x(0,z) = f_2(z)$ (at the prominence with $0 \leq z \leq h$) and $B_z(0,z) = 0$ (in the corona, $z > h$). Because of the assumed symmetry one has to solve the potential field equation only in the quadrangle $x \geq 0$ and $z \geq 0$. If we use again $\xi = x + iz$ this quadrangle can easily be mapped into the upper half-plane by the transformation $\zeta = \xi^2$. The flux function u which defines the field by $B_x = \partial u/\partial z$ and $B_z = -\partial u/\partial x$ can be obtained as the real part of $\tilde{u}$, and this complex potential is given by

$$\tilde{u}(\zeta) = \frac{1}{\pi i}\sqrt{\zeta - \sqrt{h}} \int_{\sqrt{h}}^{\infty} \frac{g(t)dt}{\sqrt{t - \sqrt{h}}(t - \zeta)} \quad (6.13)$$

where

$$g(t) = \begin{cases} u(\sqrt{t}, 0) & \text{for } t > 0 \\ u(0, \sqrt{-t}) & \text{for } -\sqrt{h} < \iota < 0 \end{cases} \quad (6.14)$$

In terms of the magnetic field one has

$$u(\sqrt{t}, 0) = \int_{\sqrt{t}}^{\sqrt{h}} f_2(\tau)d\tau$$

and
$$u(0,\sqrt{-t}) = u(0,0) + \int_0^{\sqrt{-t}} f_1(\tau)d\tau.$$

This method was applied to the field distribution shown in Figure 6.4, which is meant to represent a typical situation observed on the sun. The quantity $B_\parallel$ in Figure 6.4b is the line-of-sight component of the prominence field. Since this component is usually different from B_x we set $B_x = \beta B_\parallel$ with β as a free parameter. The solutions for two values of β are shown in Figure 6.5 These solutions are not entirely satisfactory because in the lower parts of the prominence the field lines are

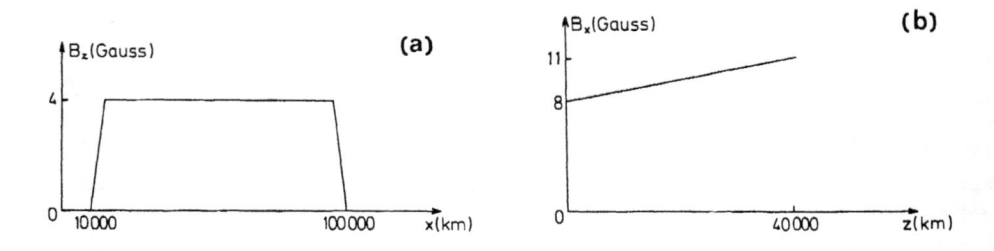

Figure 6.4 Idealisation of observed flux distribution: a) photospheric field, b) field in the prominence.

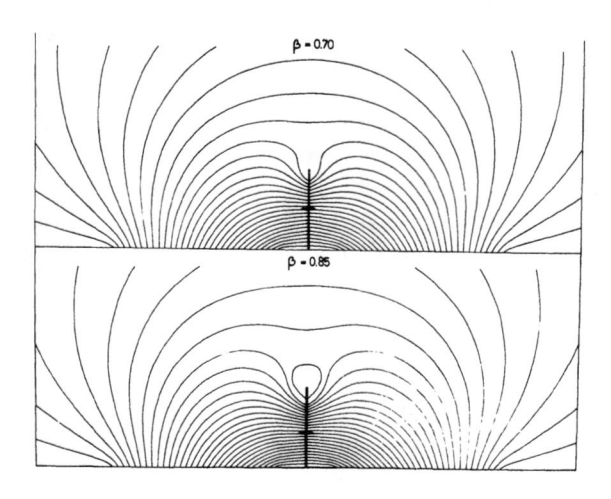

Figure 6.5 Field calculated for the flux distributions of Figure 4, where $B_x = \beta B_\parallel$ is taken.

stretched upwards and therefore give a downward Lorentz force. Only the sections above the horizontal marks in Figure 6.5 can support a prominence. There seem to be two possibilities to improve this unsatisfactory situation. One would be to vary $B_z(x,0)$ and $B_{||}(0,z)$ within the observational constraints until one gets a positive Lorentz force everywhere. The other possibility is to take into account that, with exception of the feet prominences extend only over a height interval $h_1 < z < h_2$. This gives rise to a more complicated boundary value problem (the type of boundary condition now changes three times instead of once). Such configurations are presently being studied by Démoulin, Malherbe and Priest. This modelling attempt also showed that the observed photospheric fields associated with prominences are in general not as regular as one would like them for such two-dimensional symmetric models.

6.2.1.1.2 Models with Inverse Magnetic Polarity. A different approach was taken by Kuperus and Raadu (1974). They were mainly interested in the question of what kind of magnetic field configuration can efficiently shield a prominence from the hot corona. They found that neutral magnetic sheets in which a vertical magnetic field reverses its direction are the best candidates. In such a configuration thermal conduction from the corona to the prominence has to occur perpendicular to the fields. This conductivity is extremely low and therefore the prominence is basically insulated from the corona. If such a sheet is subjected to reconnection a configuration of the type sketched in Figure 6.6a could result. This configuration can be represented by a line current, I, which is embedded in a vertical field which has opposite directions on the two sides. This field, of course, does not produce any vertical Lorentz force. Kuperus and Raadu therefore proposed that the formation of the line current will induce photospheric surface currents which prevent the field from penetrating into subphotospheric layers. Such an induction process can be simulated by a virtual line current, - I, below the photosphere. With I at the prominence height $z = h$ and - I at $z = -h$ one finds an upward Lorentz force of

$$F_L^1 = \frac{\mu I^2}{4\pi}/h \qquad (6.15)$$

The authors argue that this configuration will further develop through reconnection into something similar to what is sketched in Figure 6.6b. The new configuration is assumed to have an additional horizontal field, B_x^0, at the position of the line current. This gives rise to the downward Lorentz force

$$F_L^2 = -IB_x^0. \qquad (6.16)$$

One then has to conclude that support against gravity is possible only if the sum of these two forces is positive. The resulting field shows an inverse direction compared with the normal orientation of the Kippenhahn-Schlüter models. Whether this field has an X-type neutral point underneath the prominence as in Figure 6.1b or whether the typology is as in Figure 6.6b will depend on the field distribution in the photosphere close to the magnetic dividing line.

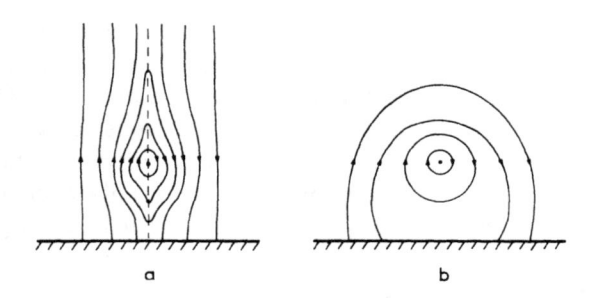

Figure 6.6 Prominence fields according to Kuperus and Raadu.

These models were later extended by van Tend and Kuperus (1978), Anzer (1984) and Anzer and Priest (1985) who included a potential field in the corona which precedes the formation of the prominence. The basic problem of these models is the question of what can drive the postulated line current I. This current must flow in the opposite direction to the one existing in the Kippenhahn-Schlüter models. In the latter the current is driven by a downward gravitational force. Therefore an upward force is required. Anzer (1984) argued that the solar wind could provide such a force. Then via reconnection one could form a line current of the right kind. This line current then will be initially pulled downward until the repulsive force due to the mirror current becomes large enough to overcome gravity. Therefore one would conclude that these prominences occur at low heights, which is contrary to the observations.

In addition prominences are not well represented by line currents. Therefore Anzer (1984) tried to extend the existing models to vertical current sheets. The following simple current distribution extending from $z = h - \delta$ to $z = h + \delta$ was used:

$$j(z) = \begin{cases} \frac{3I}{4} \frac{(h+\delta-z)(z-h+\delta)}{\delta^3} & \text{for } h - \delta < z < h + \delta \\ 0 & \text{elsewhere} \end{cases} \quad (6.17)$$

It is normalized to $\int_{h-\delta}^{h+\delta} j(z) dz = I$. The magnetic field which is produced by this current sheet can be calculated from

$$\mathbf{B}^s = -\frac{\mu}{2\pi} \int_{h-\delta}^{h+\delta} \frac{j(z') \mathbf{e}_y \times (\mathbf{r}' - \mathbf{r})}{(\mathbf{r}' - \mathbf{r})^2} dz' \quad (6.18)$$

with $\mathbf{r}' = (0, z')$. This then leads to a transverse horizontal field which varies along the sheet of the form

$$B_x^s(0, z) = -\frac{\mu}{2\pi} \frac{3I}{\delta} \left(\frac{1}{2} \frac{h-z}{\delta} + \frac{(h+\delta-z)(z-h+\delta)}{(2\delta)^2} ln \left| \frac{h+\delta-z}{h-\delta-z} \right| \right) \quad (6.19)$$

For $z > h$ one has $B_x^s > 0$ and a numerical evaluation of equation (6.19) shows a maximum of

CHAPTER 6: STRUCTURE AND EQUILIBRIUM OF PROMINENCES

$$B_x^s|_{max} \approx \frac{3.6\mu}{4\pi}\frac{I}{\delta}. \qquad (6.20)$$

at $z = z_{max}$. The induced photospheric currents can be roughly approximated by a line current, $-I$, at $z = -h$ with a field

$$B_x'(0,y) \approx -\frac{\mu}{2\pi}\frac{I}{h+z}. \qquad (6.21)$$

The estimate for B_x at $z = z_{max}$ then gives

$$B_x^s(0,z_{max}) + B_x'(0,z_{max}) > B_x^s|_{max} - \frac{\mu}{2\pi}\frac{I}{2h} \gtrsim \frac{\mu}{4\pi}\left(3.6\frac{I}{\delta} - \frac{I}{\delta}\right) > 0 \qquad (6.22)$$

since for a prominence which lies completely above the photosphere $h > \delta$ holds. This then implies that for some parts of the prominence the Lorentz force is directed downwards and equilibrium is impossible. The physical reason for such a behaviour is that an extended current sheet experiences a strong self-pinching force which is directed upwards in the lower parts of the prominence but downwards in the upper parts of the prominence. Including an additional potential field of the Kuperus-Raadu type (i.e. with $B_x^P(0,z) > 0$) makes the situation even worse. Only if $B_x^P(0,z)$ is sufficiently negative can one have detailed equilibria (these then are, however, of the Kipppenhahn-Schlüter type).

Malherbe and Priest (1983) used the method of complex functions described in the previous section to produce also configurations of the Kuperus-Raadu type. With

$$\phi(\xi) = iB_1\frac{\xi - ih}{\xi} \qquad (6.23)$$

and

$$\psi(\xi) = B_0\frac{\sqrt{p^2 + \xi^2}}{\xi} \qquad (6.24)$$

one obtains configurations as sketched in Figure 6.7. In this case the prominence is not supported from below but hangs in some kind of magnetic hammock. This

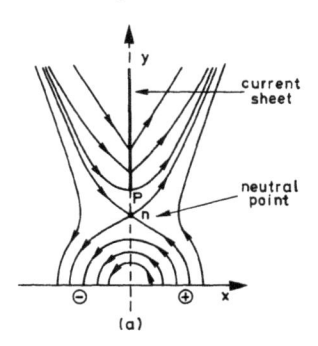

Figure 6.7 Infinite current sheet of Kuperus-Raadu type.

could only be the case if the solar wind is so strong, that it stretches the field lines to very large distances. Therefore the solar wind would have to provide the prominence support. It is not clear whether such an equilibrium can be reached and whether it would be stable. The main problem with such models is, however, that they are in conflict with the observed large field components, B_y, along the prominence. The observations show that in general B_y is much larger than B_x. Therefore to make these models work one would have to assume a stream pattern which is not radial as in the normal solar wind, but has very large horizontal components of opposite direction on either side of the prominence. This seems to be a very implausible requirement. If one uses instead of (6.23) and (6.24) the following functions

$$\phi(\xi) = -B_1 \frac{\xi - ip}{\xi(\xi - iq)} \tag{6.25}$$

and

$$\psi(\xi) = -B_0 \frac{\sqrt{(p^2 + \xi^2)(q^2 + \xi^2)}}{\xi(\xi + ih)^2} \tag{6.26}$$

one finds configuration of the type shown in Figure 6.8. They have a current sheet extending from $p \lesssim z \lesssim q$ and a line current at $z = q$. The sheet current produces an upward Lorentz force of density

$$f = \frac{2B_0 B_1}{\mu} \sqrt{\frac{z^2 - p^2}{q - z}(q + z)} \frac{z - p}{z^2(z + h)^2}. \tag{6.27}$$

This expression diverges at $z = q$, but the integrated force $F_s = \int_p^q f(z)dz$ remains finite. The line current at $z = q$ contributes a downward force of strength

$$F_l = -\frac{2\pi}{\mu} B_1^2 \frac{p(q - p)}{q^3}. \tag{6.28}$$

This means that the self-pinching encountered previously also occurs in this configuration. The problem is only shifted to the upper (singular) edge of the prominence. This seems to be a very general property of all extended Kuperus-Raadu configurations; so far at least no solution has been found. The solutions (6.25 and

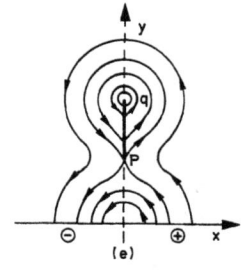

Figure 6.8 Configuration with a current sheet between p and q and an additional line current at q.

6.26) appear to have yet another unsatisfactory property: no parameter range of p, q, u has been found for which $F_s > |F_l|$; on the contrary for $h > q$ one can show that $F_s < |F_l|$ is the case. Therefore the prominence as a whole is also not in equilibrium.

All models discussed so far completely neglect the dimension along the prominence. But as discussed in the introduction the magnetic fields have in general a large y-component which results from shear motions. In the framework of 2-D potential field models such a shear can only be represented by superimposing a constant field B_y^0. This procedure preserves the equilibrium of the configurations. It is, however, unsatisfactory in several respects: since only constant B_y^0 are allowed the modelling of actual sheared fields is very restricted. And the energy associated with this field $\int_A (B_y^0)^2/2\mu dx dz$ is infinite. A third point is the problem of equilibrium for Kuperus-Raadu configurations. The only suggestion at present of how to avoid the self-pinching is to introduce a strong internal field whose pressure can balance the magnetic tension forces. But for that B_y has to vary in space.

6.2.1.1.3 Force-free fields. If one wants to incorporate the effect of magnetic shear in a more natural way one has to use force-free fields. Amari and Aly (1988) constructed such sheared fields. They assumed that the fields are force-free everywhere except at the location of the prominence. They represented the prominence by a line-current and took periodic force-free fields with constant α (where $\nabla \times \mathbf{B} = \alpha \mathbf{B}$). These fields can be represented by

$$\mathbf{B} = \nabla A \times \mathbf{e}_y + \alpha A \mathbf{e}_y. \qquad (6.29)$$

The flux function A has to fulfil the equation

$$-\Delta A = \alpha^2 A + \mu I \, \delta(\mathbf{r} - \mathbf{r}_0) \qquad (6.30)$$

with a line current I at $\mathbf{r}_0 = (0, y, h)$. The solution which is periodic in $|x| \lesssim L/2$ then is given by

$$A = B_0 L \cos\left(\frac{\pi x}{L}\right) e^{-\gamma_1 z} + \frac{\mu I}{L} \cdot \sum_{p=0}^{\infty} \frac{1}{\gamma_{2p+1}}$$
$$\cos\left((2p+1)\frac{\pi x}{L}\right) \left(e^{-\gamma_{2p+1}|z-h|} - e^{-\gamma_{2p+1}(z+h)}\right) \qquad (6.31)$$

and

$$\gamma_{2p+1} = \sqrt{(2p+1)^2 \frac{\pi^2}{L^2} - \alpha^2}. \qquad (6.32)$$

Figure 6.9 gives a rough sketch of such a field.

If the prominence mass per unit length is m then equilibrium can be reached for

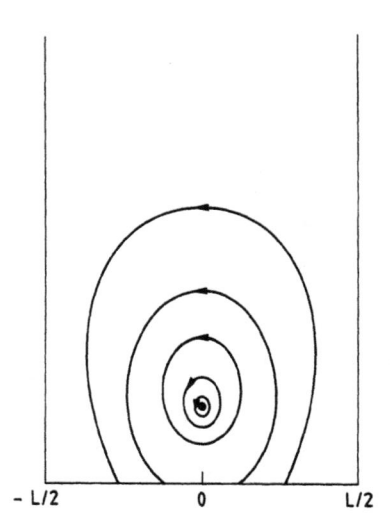

Figure 6.9 Schematic drawing of a constant α force-free field with line current.

$$\left(\frac{\mu}{L}\sum_{p=0}^{\infty} e^{-\gamma_{2p+1} h} - B_0 L \gamma_1 e^{-\gamma_1 h}\right) = mg. \tag{6.33}$$

The expression for A contains three terms. The first describes the force-free field produced by the photospheric flux distribution which is assumed to be of the form (a more general form of $B_z(x,0)$ which allows for higher harmonics in x was used by Démoulin and Priest (1988))

$$B_z(x,0) = \pi B_0 \sin\left(\frac{\pi x}{L}\right) \tag{6.34}$$

The first part of the infinite sum of equation (6.31) represents the line current at $x = 0, z = h$ and the second part gives the field of the mirror current at $z = -h$. As in the previous section the interaction of the line current with the photospheric field gives a downward pull whereas the interaction of the two line currents leads to a re pulsive force. These two competing forces are given on the left-hand side of equation (6.33). The limit $\alpha \to \pi/L$ represents infinitely sheared fields; it also implies $\gamma_1 \to 0$ and therefore the negative term in equation (6.33) tends to zero. This behaviour is different from the potential field case where one always has a finite downward pull. It is also responsible for the finding by Amari and Aly that with $m \to 0$ the prominence can reach arbitrary heights. What heights prominences will actually reach depends on the value of the physical parameters involved. The authors do not give any physical mechanism for the formation of this line current (although a twisting of large-scale flux tubes seems a possibility). Therefore no predictions about the value of I and the resulting prominence height can be made at present. There is, however, one interesting point in this context.

These solutions also allow Kippenhahn-Schlüter configurations (one simply has to replace I by -I). By inspecting equation (6.33) one finds that the equilibrium heights for Kippenhahn-Schlüter configurations are systematically larger than those of the Kuperus-Raadu type for the same values of B_0, I, α and m. As mentioned earlier the observations show that Kuperus-Raadu prominences are in general located at greater heights in the corona than Kippenhahn-Schlüter configurations. Therefore if one wants to describe both types of configuration by these force-free models one has to postulate that at least one of the above parameters has to be systematically very different in the two cases. It is not clear at present which effect would be responsible for such a duality and whether it could be confirmed by observations. So far only models with line currents have been produced, but the ultimate goal would be to obtain finite current sheets which have upward Lorentz force everywhere.

6.2.1.2 Quasi-three-dimensional models.

Real prominences show substantial variations along their horizontal axis. This then implies that more realistic models would actually have to be three-dimensional. The construction of true 3-D models is a very difficult theoretical task. So far very little has been achieved in this field. Wu and Low (1987) have gone one step in this direction. They started from special 3-D potential fields which have only horizontal componenets in the plane $x = 0$ (i.e. the location of the prominence) and superimposed the field of a 2-D current sheet placed at $x = 0$. It should be noted that these current sheets are independent of y and therefore the vertical extent of the prominence is also independent of y. The only variation occurs in the distribution of the prominence mass. It is not clear what improvements over the purely 2-D investigations can be obtained with these models and what new insight can be gained from them.

6.2.1.3 Support by Alfven waves.

Jensen (1986) proposed that prominences can be supported against gravity by Alfven waves. When the amplitudes of these waves become large enough they can be damped by nonlinear interactions and thus loose both their energy and momentum to the plasma. No detailed model exists but there seem to be several difficulties resulting from such a picture. One is the fact that the momentum transfer occurs where the heating produced by these waves is most efficient.

How can one reconcile this with the fact that prominences are very cool? The other is that the transferred momentum will be directed along the magnetic field. Since the field in prominences is predominantly horizontal the mechanism cannot produce the required lifting force.

6.2.2 INTERNAL STRUCTURE AND THERMAL EQUILIBRIUM

These two aspects are intimately connected. The density variation along the magnetic field lines is determined by pressure equilibrium which of course depends on the local temperature. On the other hand both the heating and cooling are functions of the density. Therefore the ultimate goal will be an internally consistent model of the hydrostatics and the energetics of prominences and the corona surrounding them. But so far only first steps towards this goal have been achieved. Most of the existing investigations concentrate on one or the other of these two aspects.

6.2.2.1 Hydrostatic equilibrium.

The condition for hydrostatic equilibrium is:

$$\nabla p - \rho \mathbf{g} = \frac{1}{\mu}(\nabla \times \mathbf{B}) \times \mathbf{B} \tag{6.35}$$

with

$$p = \frac{R}{\tilde{\mu}}\rho T. \tag{6.36}$$

Kippenhahn and Schlüter (1957) were the first to discuss the equilibrium structure inside prominences. They solved equation (6.35) for the simplified case that all quantities are independent of y and z. Because of the narrowness of prominences this is a reasonable approximation. Horizontal and vertical force balance then give

$$\frac{dp}{dx} = -\frac{1}{\mu}B_z\frac{dB_z}{dx} \tag{6.37}$$

and

$$\rho g = \frac{1}{\mu}B_x\frac{dB_z}{dx}. \tag{6.38}$$

Equation (6.38) leads to $p + B_z^2/2\mu = B_0^2/2\mu$, because $p \to 0$ for $|x| \to \infty$. $\nabla \cdot \mathbf{B} = 0$ leads to $B_x = const$. For constant temperature the solutions are given by

$$B_z = B_0 \tanh\left(\frac{B_0}{B_x}\frac{x}{2H}\right) \tag{6.39}$$

and

$$\rho = \frac{B_0^2}{2\mu}\frac{g}{H}\frac{1}{\cosh^2\left(\frac{B_0}{B_x}\frac{x}{2H}\right)} \tag{6.40}$$

with $H = RT/g\tilde{\mu}$. This field configuration and the density profile are shown in Figures 6.10 and 6.11. Poland and Anzer (1971) give the generalisation of equations (6.39) and (6.40) for the case that $\tilde{\mu}$ and T are prescribed functions of x. In this case one only has to replace $\tilde{\mu}x/T$ by $\int_0^x \tilde{\mu}(x')/T(x')dx'$ in the expressions for B_z and ρ. For constant temperature the width of the prominence is approximately

$$d \approx 4\frac{B_x}{B_0}H. \tag{6.41}$$

CHAPTER 6: STRUCTURE AND EQUILIBRIUM OF PROMINENCES

If $B_x/B_0 = 4$ is taken which implies an inclination angle of $15°$ between field and horizontal plane and with $H = 180 km$ we obtain $d \approx 3000 km$ which is a reasonable value for the prominence width.

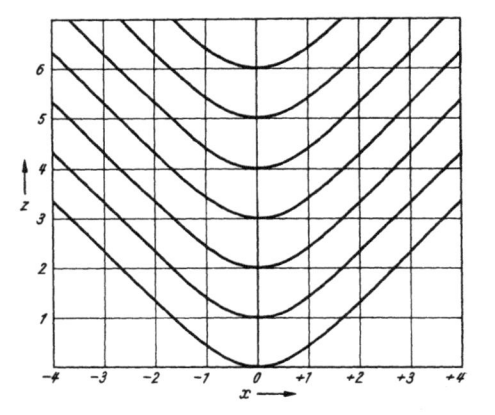

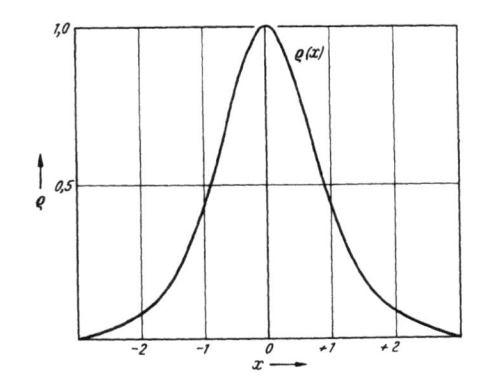

Figure 6.10 Model of the field structure inside a prominence.

Figure 6.11 Density profile for constant temperature and the field of Figure 6.10.

2-D isothermal models were constructed by Low et al. (1983). These authors allow for variations with height but assume periodicity in x. They also take a special functional form for **B**. In these highly specialised cases analytic solutions were obtained. The solutions they present have the unsatisfactory property that the prominences would have to be larger in width than in height.

Hydrostatic equilibria with varying temperature were presented by Low (1975) and Osherovich (1985). Again only very special configurations have been considered. Low (1975) studied one-dimensional configurations similar to the ones of Kippenhahn and Schlüter; the only difference is a variable temperature. To describe the field he uses the flux function F with

$$\mathbf{B} = \left(\frac{\partial F}{\partial z}, 0, -\frac{\partial F}{\partial x} \right) \tag{6.42}$$

and assumes

$$F(x,z) = \psi(x) + z. \tag{6.43}$$

For the temperature variation he arbitrarily takes the dependence

$$T = const.\psi^{2/7}. \tag{6.44}$$

In this model the temperature is prescribed as a function of ψ rather than as direct function of x as proposed by Poland and Anzer. In this case one obtains the following equation for ψ:

$$\left(\frac{d\psi}{dx}\right)^2 + A\, e^{-\gamma\psi^{5/7}} = C \tag{6.45}$$

with A, C and γ as constants. This equation then has to be solved numerically. The resulting field has the property that $\mathbf{B} \to \mathbf{B}_0 = const\cdot$ for $|x| \to \infty$ and therefore $\psi \to const|x|$. This then implies $T \to const \cdot |x|^{2/7}$ for $|x| \to \infty$ and therefore the temperature goes to infinity at large distances. Thus if one starts at a low prominence temperature these models will always reach coronal temperatures for large enough values of x. Beyond this value the solutions become unphysical and it is unclear how to match them to real coronal solutions.

Osherovich (1985) takes a very unconventional approach. He assumes special 2-dimensional fields and then calculates the pressure and density distributions which are in equilibrium with those fields. The field is described by the functions $F(x,z)$ and $G(x,z)$ with

$$\mathbf{B} = \left(\frac{\partial F}{\partial z}, G, -\frac{\partial F}{\partial x}\right). \tag{6.46}$$

He then arbitrarily chooses

$$F = F_0 \kappa z^2 e^{-\kappa(x^2+z^2)} + Hz \tag{6.47}$$

and

$$G = G_0 F \tag{6.48}$$

where F_0, κ, H and G_0 are constants. By these assumptions then $\vec{B}(x,z)$ is completely determined. The magnetohydrostatic equilibrium both along the field and perpendicular to it then gives the density ρ and the pressure p in terms of the functions F and G. This implies that $\rho(x,z)$ and $p(x,z)$ are uniquely prescribed. Then from equation (6.36) the temperature as function of x and z is obtained. The crucial question now is whether the resulting temperature field $T(x,z)$ is reasonable as far as the observations are concerned and whether the energy equation (see equation 6.49) can be fulfilled with assumptions which are physically reasonable. This investigation is still lacking and therefore it is difficult to judge the relevance of models based on this procedure.

6.2.2.2 Thermal equilibrium

To obtain complete models of quiescent prominences one has to add to the equations of MHD equilibrium (6.35) and (6.36) an energy equation of the form

$$(\mathbf{B}\cdot\nabla)\left(\frac{\kappa_\|}{B^2}\mathbf{B}\cdot\nabla T\right) = \mathcal{L} - \mathcal{H} \tag{6.49}$$

The left-hand side represents the conduction of thermal energy parallel to the magnetic field with conductivity $\kappa_\parallel$, $\mathcal{L}$ represents the radiative loss terms and $\mathcal{H}$ the heating function. The problem one faces is that both the cooling and heating mechanisms are very poorly known for the case of prominences. For an optically thin plasma one can take $\mathcal{L} = \rho^2 Q(T)$ and use Q(T) as given by Cox and Tucker (1969). This approach is reasonable for the corona surrounding the prominence but it is invalid inside prominences because they are optically thick. This then requires the solution of the complete radiative transfer problem. The situation is further complicated by the fact that the prominence is immersed in the radiation field of the solar atmosphere (although this irradiation represents a gain of energy it is usually discussed in connection with radiative loss terms). The heating of the corona is believed to be of magnetic origin, either by waves or enhanced dissipation (i.e. reconnection). The exact form of the heating function is not known, quite often it is taken as $\mathcal{H} = h\rho$ with h as an adjustable parameter. This form of the heating function is very ad hoc and its justification lies primarily in its simplicity. If the corona is heated magnetically, as generally believed, then this heating function should also depend in some way on the field. If heating occurs by Alfven waves then it should be a function of the wave amplitude, δB. In the case of magnetic dissipation it should depend on the current density (or B/l, where l is the typical width of the current sheets involved) and also on the effective resistivity, η_{eff}, which will be a function of the temperature. No detailed models for these heating processes exist so far. An additional complication with wave heating is that the waves loose their energy as they travel along the field and thus the heating will decrease along the field. Some of these aspects could play a role also for the process of prominence formation. This was discussed in the investigation of prominence condensation by Hood and Anzer (1988). The situation becomes even more complex for established prominences. In this case the density changes almost discontinuity from the corona to the prominence. This means that the waves will be reflected giving a reduced wave heating inside the prominence. On the other hand the low temperatures might increase the reconnection rates by a large factor. Because of all these uncertainties the models which we shall now discuss have to be considered with some caution.

Poland and Anzer (1971) developed a model in which they divided the prominence into two parts: the interior of the prominence and the transition region between the cool prominence and the hot corona. They assumed that in both regions mechanical heating can be neglected; i.e. $\mathcal{H} = 0$ was chosen. For the inner part they took a gas of constant temperature and a density which decreases exponentially outward. For the radiative transfer model they used a three-level hydrogen atom with two bound states and the continuum. The Lyman continuum turns out to be saturated in the prominence whereas the Balmer continuum is optically thin. The prominence is irradiated from both sides by the photospheric, chromospheric and coronal radiation field. This radiation field falls off exponentially inside the prominence

$$J_{\nu ext}(\tau_\nu) = J_{\nu \mathrm{imp}} e^{-\tau_\nu} \qquad (6.50)$$

where $J_{\nu imp}$ is the impressed field and τ_ν the optical depth measured from the prominence surface. The transfer calculations were performed for a prominence of constant temperature (typically 6000 K) and slab geometry. These calculations gave the degree of ionisation, the population density of the $n = 2$ level and the energy flow in the Lyman continuum as functions of depth. The H_α transfer was not treated in detail. Instead the H_α losses were estimated by

$$\epsilon_{H_\alpha} = \bar{N}_2 \, C_{23} \, h\nu \, d/2 \qquad (6.51)$$

where $\bar{N}_2$ is the average $n = 2$ density and C_{23} the collision rate. Some test calculations which included the $n = 3$ level showed that the estimates resulting from equation (6.51) were quite good. The computations for T = 6000 K and N_H between 10^{17} and $4 \times 10^{18} m^{-3}$ gave a gain of energy in the Lyman continuum of about $90 W m^{-2}$. From $\bar{N}_2 \approx 5 \times 10^{10} m^{-3}$ and $\bar{N}_2 d \approx 3 \times 10^{17} m^{-2}$ one obtains $\epsilon_{H_\alpha} \approx 2 \times 10^2 W m^{-2}$; T = 7500 K would result in $\epsilon_{H_\alpha} \approx 4 \times 10^2 W m^{-2}$. The other main contributions to the radiative losses are the H- and K-line of calcium for which estimates gave $\approx 30 W m^{-2}$. Therefore the net radiative losses from these model prominences amount to approximately $1.3 \times 10^2 W m^{-2}$.

The transition from the cool prominence to the hot corona is expected to take place in a thin sheath-like region surrounding the prominence. We assume that the magnetic field has no divergence in this region and also that no energy sinks or sources are present there. Then from equation (6.49) we obtain

$$\frac{d}{ds}\left(\kappa' T^{5/2} \frac{dT}{ds}\right) = 0 \qquad (6.52)$$

where s is the coordinate along the field and κ' the constant determining the thermal conductivity. From equation (6.52) one obtains the thermal flux:

$$\begin{aligned} F = \kappa' T^{5/2} \frac{dT}{ds} &= \frac{2}{7}\kappa' \frac{T_{cor}^{7/2} - T_{pr}^{7/2}}{\delta s} \\ &\approx \frac{2}{7}\kappa' \frac{T_{cor}^{7/2}}{\delta s}, \end{aligned} \qquad (6.53)$$

$\delta s = \delta x / \sin \gamma$ with δx the width of the transition region and γ the angle between field and prominence. Taking $\gamma = 15°$ as suggested by the observations one finds for $\delta x = 1000 km$ and $T_{cor} = 5 \times 10^5 K$ $F = 50 W m^{-2}$ and for $\delta x = 500 km$ and $T_{cor} = 10^6 K$ $F = 1.3 \times 10^3 W m^{-2}$; as a representative situation we choose $\delta x = 500 km$ and $T_{cor} = 5 \times 10^5 K$ which results in $F = 100 W m^{-2}$. This amount of energy inflow can be radiated away and therefore thermal equilibrium can be achieved in this model. Although overall energy balance is possible there remains one big problem: to radiate the energy efficiently away it first has to be conducted to regions in the prominence where the optical depth $\tau \approx 1$. But in these cool regions the thermal conduction is very low and thus the energy can only be transported over short distances. This problem has not been solved yet. One possibility would be that a strongly filamentary structure might allow the thermal energy to penetrate deep into the prominence.

Heasley and Mihalas (1976) studied the radiative transfer problem in prominences in much greater detail. They assumed that the prominence consists of 90% hydrogen and 10% helium. Hydrogen was treated as an atom with 5 bound levels and the continuum, He^0 and He^+ both were taken as 2-level systems together with the continuum. Slab geometry was used. As in the investigation by Poland and Anzer the irradiation of the prominence by the adjacent atmosphere was taken into account. The authors studied a variety of models. Some of these have prescribed constant temperatures (as in Poland and Anzer). Because these models are not entirely in equilibrium the authors then procede to construct models which are in pure radiative equilibrium. They set the column density (ρd) to $2 \times 10^{-3} kgm^{-2}$ and investigate prominences of different widths ranging from 1000 km to 11000 km. These models give surface temperatures between 7000 and 8000 K and central temperatures of about 4600 K. These central temperatures are far too low. The models can also be tested against the observations using the Balmer decrements. One observes

$$I(H_\alpha) : I(H_\beta) : I(H_\gamma) \approx 4.5 : 1 : 0.45 \quad (6.54)$$

whereas the models predict ratios of 7.2 : 1 : 0.27 which is completely unacceptable.

One can improve the situation if one postulates a highly filamented prominence structure which allows the Lyman continuum radiation to penetrate deep into the prominence interior. Observations indicate that this is actually the case for prominences. The authors use the filling factor γ to describe this clumpiness in a simplistic way. Computations for $\gamma = 0.1$ and $\gamma = 0.2$ were carried out. This implies a drastic change of the original picture: e.g. $\gamma = 0.1$ means that the density in the threads is $10^{18} m^{-3}$ and more and only one tenth of the prominence volume is actually filled with dense material. Such models give better agreement with the observations. For $\bar{\rho} d = 2 \times 10^{-3} kgm^{-2}$, $d = 3800 km$ and $\gamma = 0.1$ one finds 7400 K at the surface and 5400 K in the centre of the prominence, which is still somewhat too low. The resulting temperature structure is shown in Figure 6.12 with the column density m as abscissa; m is defined as

$$m = \int_x^{d/2} \bar{\rho} dx \quad (6.55)$$

where $\bar{\rho}$ is the mean density of the filamentary prominence. The Balmer decrements are now 3.8 : 1 : 03 which is also in better agreement with the observations. These models go in the right direction, but they should be considered with caution: they solve the radiative transfer for a plane parallel slab geometry and at the same time a very inhomogeneous structure is assumed. Because of these mutually contradicting assumptions it is difficult to judge how reliable the obtained results are. Another

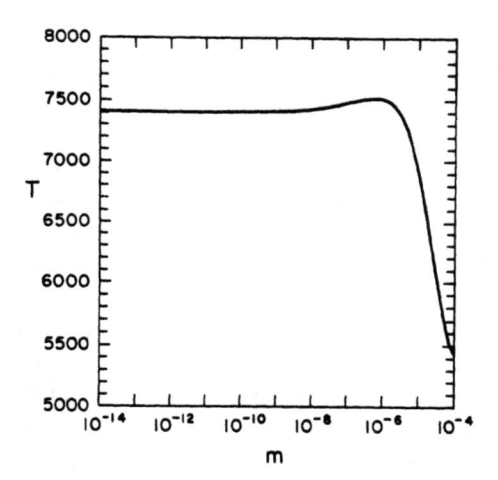

Figure 6.12 Temperature structure with "diffusive" penetration.

series of models are those with internal heating. The heating function is taken as

$$\mathcal{H} = h\rho \tag{6.56}$$

with $h = 1 - 6 \times 10^5 W kg^{-1}$. Such models give an even better agreement with the observations. If one takes $h = 1.5 \times 10^5 W kg^{-1}$ then the temperature goes from 8300 K at the surface to 7000 K in the center and the Balmer decrements become 4.3 : 1 : 0.3. The heating of a column of unit cross section is $\mathcal{H}d = 3 \times 10^2 W m^{-2}$. It is interesting to compare this to the thermal influx taken by Poland and Anzer of $2F = 2 \times 10^2 W m^{-2}$. One may conclude from this that heating by conduction, or waves or gravitational energy release (or a combination of them) can produce models which are in agreement with the observations.

Lerche and Low (1977) produced prominence models by arbitrarily setting the difference between the loss term and the heating function in equation (6.49) proportional to the density. They assumed

$$\mathcal{L} - \mathcal{H} = \Lambda \rho \tag{6.57}$$

with Λ being a free parameter, and also set $\kappa_{,,} = K_0 T$. Both these assumptions are not very realistic. Therefore it is hard to judge in what sense the resulting models can describe the physics of prominences. Milne et al. (1979) solved the full energy equation (6.49) but only for special simple heating and cooling functions. They also neglected the effects of irradiation from the surroundings. They also used 1-D slab geometries. The divergence of B in the thermal conduction term was left out but the shielding by magnetic shear was included. For the radiative cooling they took the optically thin approximation of Cox and Tucker (1969). These can be represented by

$$\mathcal{L} = C\rho^2 T^\alpha \tag{6.58}$$

where C and α were taken as the constants and exponents given by Hildner (1974), and shown in Table 6.1. For the heating they took

$$\mathcal{H} = h\rho. \tag{6.59}$$

Table 6.1: Parameters for cooling functions $\mathcal{L} = C\rho^2 T^\alpha$ (in MKS units).

T	C	α
$T < 1.5 \times 10^4$	4.9×10^{-13}	7.4
$1.5 \times 10^4 < T < 8 \times 10^4$	1.2×10^{11}	1.8
$8 \times 10^4 < T < 3 \times 10^5$	8.0×10^{19}	0
$3 \times 10^5 < T < 8 \times 10^5$	3.9×10^{33}	-2.5
$8 \times 10^5 < T$	5.5×10^{24}	-1.0

The outer boundary was placed at $x_1 = RT_1/g$ with $T_1 = 2 \times 10^6 K$; and a coronal density of $n_1 = 10^{14} m^{-3}$ was assumed. At this boundary they set $\mathcal{H} = \mathcal{L}$ leading to $h = C\rho_1 T_1^\alpha$. This choice is somewhat arbitrary, since it is not clear why heating should exactly balance cooling at this specific temperature. The magnetic field $\mathbf{B} = (B_0, B_y, B_z(x))$ can be characterized by the parameters $\beta = 2\mu p_1/B_0^2$ and $\tan\varphi = B_y/B_0$. These parameters together with the above boundary conditions and symmetry about $x = 0$ then determine the solutions completely. The equations have to be solved numerically. The authors investigated a wide range of parameters and found that solutions with a low temperature at $x = 0$, called "cool solutions" can only be obtained if either $\beta \approx 0.6$ or $\varphi \approx 80°$. The first possibility is ruled out by the fact that the magnetic field in prominences is normally 5 Gauss or more giving $\beta \lesssim 3 \times 10^{-2}$. The alternative of taking $\varphi \approx 80°$ seems possible, although the observations indicate $\varphi \approx 60° - 75°$, but it seems somewhat implausible that condensations should be restricted to $\varphi \approx 80° \pm 2°$. More serious is that the temperature and density profiles are not in agreement with the observations. The full line in Figure 6.13 shows such a profile. It represents a central temperature of only 5000 K and a ridiculously small width of 25 km!

This situation has lead the authors to study modifications of the heating and cooling functions. It seems quite plausible that the heating inside prominence will be much below that given by equation (6.58). This effect was simulated by setting (somewhat arbitrarily) $\mathcal{H} = 0$ for $T < 200000 K$. Results for this case are also shown in Figure 6.13. The prominence is now somewhat wider (≈ 60km) but extremely cool ($T_0 \approx 400$ K !). The only positive aspect is that condensations with lower β (of ≈ 0.2) are now possible. The next step was to change the cooling function. Since prominences are in reality optically thick the radiative losses from the inner parts will be greatly reduced. This has been taken into account by

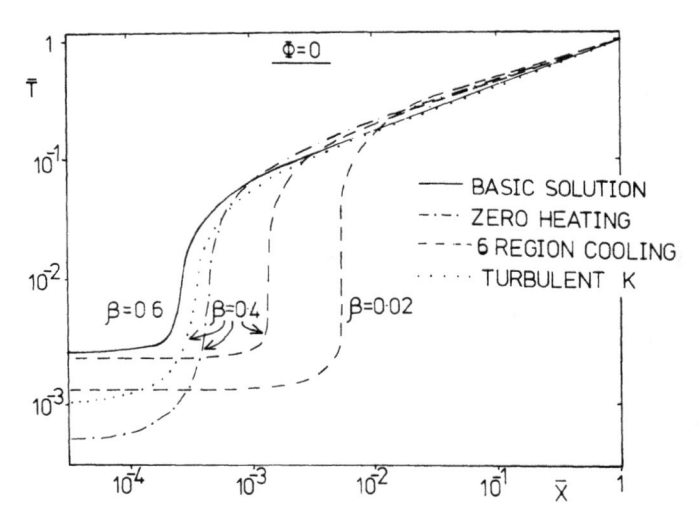

Figure 6.13 Temperature profiles for different heating and cooling functions and orientations of the field.

changing the exponent in the cooling functions. The authors take $\alpha = 17.4$ (instead of 7.4) for $T < 10000K$; this gives a reduction of 10^3 at $T = 5000K$. Curves for reduced heating and cooling are also presented in Figure 6.13. Now values of $\beta = 0.02$ are allowed and the width rises to 700 km, but the central temperature remains much too low.

Although the modifications used do not give configurations which are completely satisfactory they are quite valuable because they show quite clearly the trends and magnitudes resulting from the changes in the physics. However, it seems unlikely that one can obtain much more realistic models in this way. The reason is that the reduction of the heating and cooling is taken as function of temperature only. But one expects that in reality these effects will depend primarily on the integrated column density.

No one has so far developed models of internal equilibrium and energy balance for magnetic field configurations of the Kuperus-Raadu type. One reason is certainly that up to now no current sheet equilibria of this type have been found. The other is the belief that locally there should be little difference between Kuperus-Raadu and Kippenhahn-Schlüter models. This might be true for the hydrostatic aspect of the problem but the energetics will be very different in the two cases. The magnetic fields in Kuperus-Raadu prominences are either closed (2-D case) or helical (3-D case). This has far reaching consequences for the thermal conduction and the heating. In such configurations the prominence is basically shielded from the hot corona and no heat is conducted into the prominence. This is the original motivation of Kuperus and Raadu to consider such configurations. A similar argument holds for the magnetic heating which has its source in the motion of the photospheric foot points of the field lines. This then means that helical fields will be very inefficient in heating the prominences. One may conclude that this heating should be negligible compared to that which takes place in Kippenhahn-Schlüter prominences. This then implies that without some other kind of energy supply Kuperus-Raadu configurations would not be in thermal equilibrium since there will be radiative losses which are not balanced. The situation becomes different for the case that irradiation by the Lyman continuum becomes sufficiently large. The models of Heasley and Mihalas show that under these circumstances thermal equilibrium is possible. But the temperature is much too low. By introducing diffusive penetration one can raise the temperatures in their models up to 5400 K. This has to be compared to their models with mechanical heating which give much higher temperatures. This then leads one to the following prediction: if the effects of the Lyman continuum are small Kuperus-Raadu prominences will not be in thermal equilibrium and thus cool down continuously. And even if the Lyman continuum is important the Kuperus-Raadu configurations should have systematically 1500 K to 2000 K lower temperatures than the Kippenhahn-Schlüter prominences.

6.3 Concluding Remarks.

We have tried to show in this chapter the wide range of attempts to model quiescent prominences. The results so far have not been entirely satisfactory. The biggest mystery is still why there are these two distinct classes of magnetic field topology. The basic question in this context is what are the physical conditions which will give rise to prominences with Normal Polarity and which ones will lead to those with Inverse Polarity. In the case of configurations with Normal Polarity the prominence currents follow naturally from the gravitational force of the dense heavy material. For these, equilibrium models of 2-dimensional thin sheets exist. What is still lacking in this case is a consistent incorporation of the observed large B_y-fields. For this one needs models with a sheet current and force-free volume currents outside the prominence. For models with Inverse Polarity the situation is less satisfactory. It is not clear by which process the equilibrium currents are produced. Also the effect of self-pinching becomes crucial and this has led to the fact that no sheet equilibria have been found so far. Here too, the solution could lie in the B_y-fields. If there is a strong internal y-component of the field its pressure could prevent the prominences from collapsing.

The task becomes even more difficult when one tries to model the actual 3-dimensional structures. The most striking 3-dimensional feature is the existence of a sharp lower boundary. This boundary is shaped like arches with feet which in general reach down to the solar surface but occasionally seem to terminate at somewhat larger heights. This curved lower boundary has probably to do with the magnetic field structure. But about its physical reason one can only speculate. One possibility is that underneath the arches the divergence of the field is sufficiently small. Then the effects of thermal conduction and wave guiding could be so strong that they prevent prominence condensation. The other reason could be that the field at these locations - where it certainly is strong enough - has a geometry which does not allow stable prominence support. A possible way to model these 3-dimensional structures could be by using force-free constant - α, fields which are periodic in the x- and y-directions. Here L_y would have to be the length of the periodic arches and L_x should represent the size of the bipolar region. In addition to this one observes that prominences are by no means uniform but consist of many vertical threads of high density. It has been noticed that this strong variation in density is not accompanied by a similar variation in the magnetic field strength. This, however, is not so surprising when one looks at the equilibrium conditions. In the y-direction the total pressure has to be constant and since $\beta \ll 1$ even large density fluctuations will only produce small changes in B. For the vertical equilibrium gravity has to be balanced by the tension force $B_x B_z / \mu$. A variation in density will therefore lead to a variation in B_z. But since B_z is usually small in prominences one would not expect that such variations will be easily detected. From this one may conclude that the observations do not contradict each other. But one must also admit that the physical reason for the formation and maintenance of the threads has not been found.

The modelling of the internal structure is also far from complete. The main uncertainty lies in the effect the irradiation, predominantly in the Lyman continuum, has on the thermal balance. What is also urgently needed is an understanding and quantitative description of the heating processes taking place inside the prominence.

All our models are static. This seems to be a reasonable first approximation because the velocities observed are much smaller than any dynamical velocity of the system. Moreover present observations do not allow us to distinguish between configurations with small downward motions and no motions at all.

On the other hand oscillatory horizontal motions seem to be present in many prominences. These oscillations could become a useful tool to probe prominences in the near future. One can imagine that observations leading to a k-ω-diagram together with static equilibrium models could eventually determine physical parameters like the magnetic field and the mass distribution. This chapter can be concluded by the remark that we have some basic concepts of quiescent prominences but there is still a lot of room for imagination and ingenuity for developing better prominence models.

CHAPTER 7

STABILITY AND ERUPTION OF PROMINENCES

A. W. Hood
Mathematical Sciences Department
The University
St Andrews KY16 9SS
Scotland

7.1 Introduction

Stability theory has several important roles to play in describing solar prominences. Firstly, the amazing stability of the large quiescent prominences and the smaller active-region filaments must be explained and, secondly, stability theory may predict the sudden eruption of a prominence once critical conditions are reached. Thirdly, the fine structure observed in prominences may be due to localised instabilities that do not disrupt the overall magnetic field pattern. Finally, the formation and long lifetimes suggest that thermal instabilities must also be considered.

The need to explain prominence stability can be seen by a simple time-scale analysis. Taking a typical field strength of $10\,G$ and coronal and prominence densities of $10^{15}\,m^{-3}$ ($10^9\,cm^{-3}$) and $10^{17}\,m^{-3}$ ($10^{11}\,cm^{-3}$) then the typical coronal and prominence Alfvén velocities are $700\,kms^{-1}$ and $70\,kms^{-1}$, respectively. For typical prominence lengths, widths and heights of $10^5\,km$, $5 \times 10^3\,km$ and $5 \times 10^4\,km$ the relevant prominence Alfvén time-scales, τ_A, are shown in Table 1.

	τ_A	τ
$10^5\,km$	24 mins	
$5 \times 10^3\,km$	1.2 mins	35 days
$5 \times 10^4\,km$	12 mins	

Table 1 Typical prominence Alfvén time-scales for typical prominence lengths. The shortest tearing mode time-scale is also shown

Thus, it can be seen that prominences must be stable to ideal MHD disturbances since their lifetimes are substantially longer than the Alfvén time-scale. Resistive instabilities grow on a hybrid time-scale, for example, $\tau_A^{2/5}\tau_d^{3/5}$, where τ_A is the Alfvén time-scale and $\tau_d = l^2/\eta$ is the diffusive time-scale. l is the length-scale for which the resistive mode is effective. From Priest (1982), $\eta = 10^9 T^{-3/2}\,m^2\,s^{-1}$, where the Coulomb logarithm has been taken as 20. Taking $T = 7 \times 10^3\,K$, the tearing mode time-scale, τ, is given in Table 1, where the tearing length-scale l has been taken as half the prominence value. Obviously, this time is reduced if anomalous resistivity is used and if the length-scales for tearing are some fraction of the prominence values, for example one tenth.

The conclusion then is that quiescent prominences are globally stable to both ideal and resistive

MHD disturbances. However, during the eruption stage an instability evolves and the prominence disappears within a matter of minutes to hours. Thus the eruption can be described in terms of MHD phenomena.

The study of the stability of magnetic fields is sufficiently complicated that the basic state is usually assumed static. This neglect of equilibrium flows is reasonable since the observed flows are usually the order of a few kms^{-1} and so are highly sub-Alfvénic. An expansion in terms of the Alfvénic mach number produces the static equilibria equations as the leading order approximation.

In Section 2 a brief description of the basic MHD instabilities is presented and is followed by a resumé of the methods of solution in Section 3. In the next section, the strong stabilising influence of the dense photosphere is discussed. These results are then applied to coronal and prominence magnetic fields in Section 5 and thermal stability is discussed in Section 6. Section 7 describes resistive instabilities and their influence on prominence stability and Section 8 describes a simple model for a prominence eruption. Concluding remarks are presented in Section 9.

7.2 Description of MHD Instabilities

It is possible to understand the main MHD instabilities by considering simple equilibrium magnetic fields. The following physical descriptions use the fact that the magnetic field produces a pressure force when compressed and a tension force when the field lines are bent. The tension, in general, produces a strong restoring force that stabilises any instability and so the most dangerous disturbances are those that minimise the perturbed tension.

Consider the linear pinch, in which a *uniform* axial current (j_0 per unit volume) flows along a plasma cylinder. An azimuthal magnetic field, B_θ, is produced by the current and the $\mathbf{j} \times \mathbf{B}$ force acts radially inwards to compress and contain the plasma. Equilibrium is obtained by balancing this with an outward pressure gradient. Thus, the current is given by

$$\mathbf{j} = \begin{cases} j_0 \hat{z} = (rB_\theta)'/\mu r \hat{z} & r < r_0, \\ 0 & r > r_0, \end{cases}$$

and so

$$B_\theta = \begin{cases} \mu j_0 r/2, & r < r_0, \\ B_1/r & r > r_0, \end{cases}$$

$$p = \begin{cases} p_0 - \mu j_0^2 r^2/4 & r < r_0, \\ 0 & r > r_0, \end{cases}$$

where $p_0 = \mu j_0^2 r_0^2/4$ is the pressure on the axis and $B_1 = \mu j_0 r_0^2/2$ from continuity of total pressure at $r = r_0$. It is assumed that the total current flowing in the plasma remains constant.

Firstly, consider the *sausage mode instability*. In this case the plasma is disturbed as shown in Figure 7.1. At the constriction the radius is reduced but since the total current is conserved the current density, and hence B_θ, is enhanced resulting in a local increase in magnetic pressure. At the bulges the opposite effect occurs and there is a decrease in magnetic pressure. Thus a total pressure gradient is created that forces the plasma away from the compressed region into the extended region. Since the reduced gas pressure can no longer provide a balance for the magnetic forces, the disturbance continues to grow. The sausage mode instability can be stabilised by imposing an axial field, B_z, inside the plasma. This field provides two stabilising effects through both an enhancement of magnetic pressure inside the plasma column at the constriction, which counter-balances the enhanced external pressure, and also the introduction of magnetic tension terms.

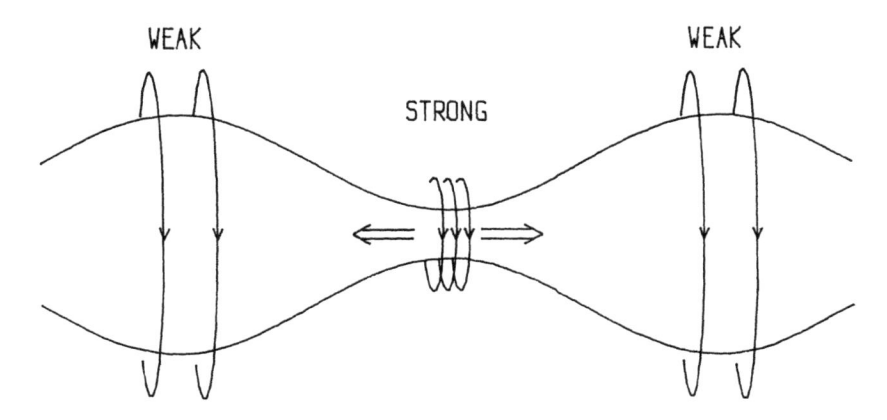

CHAPTER 7: STABILITY AND ERUPTION OF PROMINENCES   169

Figure 7.1 The sausage mode instability

Secondly, consider the *kink mode instability*. If the plasma column is disturbed as shown in Figure 7.2, then a similar argument can be used to show that the magnetic pressure is enhanced on the inside of the curve whereas it is reduced on the outside. In this case the plasma pressure and the column radius can remain unchanged, so there is no force to balance the magnetic pressure gradient and the disturbance continues to grow. Again an axial magnetic field can stabilise this lateral kink but the plasma can still be unstable to a helical mode.

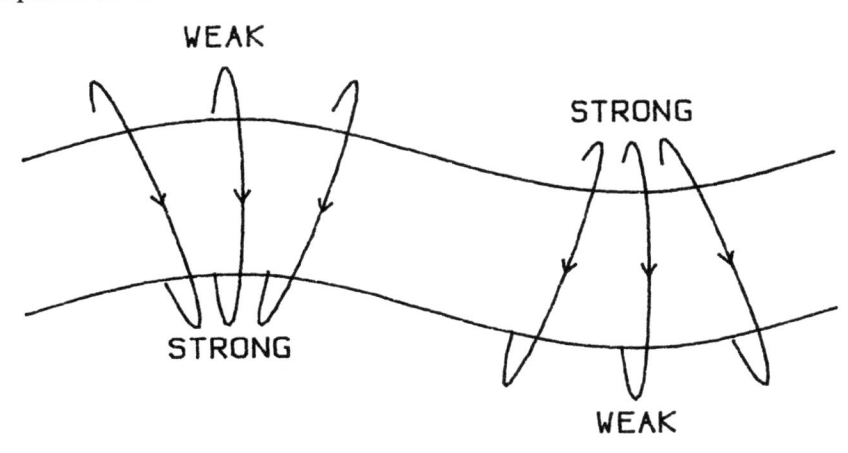

Figure 7.2 The kink mode instability

If the plasma column is cylindrically symmetric, then disturbances can be decomposed into individual Fourier modes of the form

$$f(r, \theta, z) = f(r) e^{i(m\theta + kz)},$$

where the azimuthal wavenumber m is used to describe the instability. The sausage mode is given by $m = 0$ and the (helical) kink mode by $m = 1$. Modes with larger values of m are called *localised* modes. While such a simple classification of the various instabilities may not be possible for more general two-dimensional fields, the kink mode is characterised by the fact that it is the only mode for which the axis of symmetry is displaced. Normally, the kink mode is the most dangerous in that it has a lower threshold and is global in nature as opposed to the higher threshold and more localised nature of the high m modes. In laboratory situations, the kink mode can be stabilised by a perfectly conducting shell but this stabilising feature is not present in astrophysical situations.

The *Rayleigh-Taylor* instability occurs when a high density plasma overlies a lighter one and it is driven by the effect of gravity. Thin 'fingers' of heavier fluid fall while larger 'bubbles' of lighter fluid rise. A simple proof of the instability condition can be obtained by considering the situation where the two fluids are separated by a sharp interface and where ρ_+ and ρ_- are the upper and lower plasma densities, respectively. Consider a sinusoidal disturbance of the interface, such that the area of the lower fluid now raised above the equilibrium level is the same as that of the upper fluid lowered below the equilibrium level. Then, the change in the potential energy is simply $-\rho_+ g A d$ for the upper fluid and $+\rho_- g A d$ for the lower fluid, where g is the gravitational acceleration, A is the area and $d/2$ is the distance of the centre of mass of the perturbed regions above and below the interface. The total change in potential energy is given by

$$\delta W = (\rho_- - \rho_+) g A d.$$

Therefore, when $\rho_+ > \rho_-$, potential energy is reduced by this disturbance and this energy is free to drive motions that continue the growth of the perturbation.

These simple intuitive descriptions of MHD instabilities can help with a basic understanding of the modes but a more rigorous approach is necessary for more complicated equilibria and for studying the conditions in the solar corona and prominences.

7.3 Methods of Solution

7.3.1 NORMAL MODES

The most obvious method for studying the stability of a given equilibrium magnetic field is to linearise about this state and generate a set of linear equations. These equations can then be solved by looking for normal mode solutions. Thus, setting

$$\begin{aligned} p &= p_0(\mathbf{r}) + p_1(\mathbf{r}, t), \\ \rho &= \rho_0(\mathbf{r}) + \rho_1(\mathbf{r}, t), \\ \mathbf{B} &= \mathbf{B}_0(\mathbf{r}) + \mathbf{B}_1(\mathbf{r}, t), \\ \mathbf{v} &= \mathbf{v}_1(\mathbf{r}, t), \\ \mathbf{j} &= \mathbf{j}_0(\mathbf{r}) + \mathbf{j}_1(\mathbf{r}, t), \end{aligned} \quad (7.1)$$

substituting into the MHD equations (1.1)-(1.6) and neglecting products of perturbed quantities produces the linearised system of equations

$$\frac{\partial \rho_1}{\partial t} + \nabla \cdot (\rho_0 \mathbf{v}_1) = 0. \quad (7.2)$$

$$\rho_0 \frac{\partial \mathbf{v}_1}{\partial t} = -\nabla p_1 + (\nabla \times \mathbf{B}_1) \times \mathbf{B}_0/\mu + (\nabla \times \mathbf{B}_0) \times \mathbf{B}_1/\mu + \rho_1 g, \quad (7.3)$$

CHAPTER 7: STABILITY AND ERUPTION OF PROMINENCES 171

$$\frac{\partial p_1}{\partial t} + \mathbf{v}_1 \cdot \nabla p_0 = -\gamma p_0 \nabla \cdot \mathbf{v}_1, \qquad (7.4)$$

$$\frac{\partial \mathbf{B}_1}{\partial t} = \nabla \times (\mathbf{v}_1 \times \mathbf{B}_0) + \eta \nabla^2 \mathbf{B}_1, \qquad (7.5)$$

$$\nabla \cdot \mathbf{B}_1 = 0. \qquad (7.6)$$

It is assumed, for the moment, that the plasma evolves adiabatically and so thermal and resistive effects are neglected. This is justifiable when the MHD time-scales are substantially shorter than the radiative time-scales. The perturbed velocity is related to the Lagrangian displacement vector by

$$\mathbf{v}_1 = \frac{\partial \boldsymbol{\xi}}{\partial t}. \qquad (7.7)$$

It is possible to reduce equations (7.2)-(7.7) to the linearised equation of motion

$$\rho_0 \frac{\partial^2 \boldsymbol{\xi}}{\partial t^2} = \mathbf{F}(\boldsymbol{\xi}), \qquad (7.8)$$

where the force function is

$$\mathbf{F}(\boldsymbol{\xi}) = \nabla(\boldsymbol{\xi} \cdot \nabla p_0 + \gamma p_0 \nabla \cdot \boldsymbol{\xi}) - \nabla \cdot (\rho_0 \boldsymbol{\xi})g + (\nabla \times \nabla \times (\boldsymbol{\xi} \times \mathbf{B}_0)) \times \mathbf{B}_0/\mu$$
$$+ (\nabla \times \mathbf{B}_0) \times \nabla \times (\boldsymbol{\xi} \times \mathbf{B}_0)/\mu \qquad (7.9)$$

Equations (7.8) and (7.9) are then the starting point for studying the *ideal MHD stability* of coronal magnetic fields, in general, and prominence fields, in particular. To use the *normal mode method*, the coronal displacement takes the form $\boldsymbol{\xi} e^{i\omega t}$ so that (7.8) reduces to the eigenvalue problem

$$-\rho_0 \omega^2 \boldsymbol{\xi} = \mathbf{F}(\boldsymbol{\xi}). \qquad (7.10)$$

This is, in general, a complicated set of equations to solve and analytical progress is only possible for the simplest of equilibria. The spectrum for ω^2 exhibits a discrete part and, in many cases, a continuous part. The continuous part of the spectrum consists of eigenfunctions that exhibit a logarithmic singularity and care must be taken in such cases (Goedbloed, 1983; Roberts, 1984). However, it is usual that the continuous spectrum lies in the stable region with $\omega^2 > 0$. The point of marginal stability, $\omega^2 = 0$, normally indicates the lower limit of the continuous spectrum and only discrete modes exist in the unstable region with $\omega^2 < 0$. It is not yet obvious how these results, for an infinite plasma, are effected by the solar coronal boundary conditions that restrict the length of the field lines to that portion lying above the photosphere. To avoid possible complications with continuous spectra, it is normal to obtain the threshold for instability by approaching marginal stability from the unstable region.

As an example of the normal modes method, consider the Rayleigh-Taylor instability described above. Following Wesson (1981), the equilibrium is

$$\rho = \begin{cases} \rho_+ & z > 0, \\ \rho_- & z < 0, \end{cases}$$

where ρ_+ and ρ_- are constants. Since the fluid is uniform in x and y, all perturbed variables may be expressed in terms of a single Fourier component

$$f(z) e^{i(kx - \omega t)}$$

just in terms of x by selecting appropriate coordinate axes. Assuming the plasma is incompressible, so that $\nabla \cdot \mathbf{v}_1 = 0$ is the equation of state instead of (7.4), and substituting into the linearised equation of motion (7.3) and mass continuity (7.2) gives

$$-i\omega \rho_1 = -v_z \frac{d\rho_0}{dz}, \tag{7.11}$$

$$-\rho_0 i\omega v_x = -ikp_1, \tag{7.12}$$

$$-\rho_0 i\omega v_z = -\frac{dp_1}{dz} - \rho_1 g. \tag{7.13}$$

A single equation for v_z may be derived, namely,

$$\frac{d}{dz}\left(\rho_0 \frac{dv_z}{dz}\right) - k^2\left(\rho_0 + \frac{g}{\omega^2}\frac{d\rho_0}{dz}\right)v_z = 0. \tag{7.14}$$

Since the densities are constant inside each fluid, continuity of normal velocity across the interface gives the solutions as

$$v_z = v_z(0) e^{\mp kz},$$

so that $v \to 0$ as $z \to \pm\infty$. The other matching condition is obtained by integrating (7.14) across the interface

$$\left[\rho_0 \frac{dv_z}{dz} - \frac{k^2 g}{\omega^2}\rho_0 v_z\right]_{z=0_+} = \left[\rho_0 \frac{dv_z}{dz} - \frac{k^2 g}{\omega^2}\rho_0 v_z\right]_{z=0_-}. \tag{7.15}$$

Substituting the solutions from each region into (7.15) produces the dispersion relation,

$$\omega^2 = kg\frac{(\rho_- - \rho_+)}{(\rho_- + \rho_+)}. \tag{7.16}$$

In a similar manner, a uniform, horizontal magnetic field can be included and the dispersion relation is then modified by the inclusion of a surface Alfvén wave, so that

$$\omega^2 = kg\frac{(\rho_- - \rho_+)}{(\rho_- + \rho_+)} + \frac{2(\mathbf{k}\cdot\mathbf{B}_0)^2}{\mu(\rho_- + \rho_+)}, \tag{7.17}$$

where now $k^2 = k_x^2 + k_y^2$ and $\mathbf{k} = (k_x, k_y, 0)$. The magnetic field stabilises the instability except when the displacements are uniform along the field ($\mathbf{k}\cdot\mathbf{B}_0 = 0$) so that the field lines remain straight and no tension forces are produced.

7.3.2 ENERGY METHOD

When the equilibrium is complicated, it is often easier to obtain information about the plasma's stability by using the energy method of Bernstein *et al* (1958). Taking the scalar product of (7.10) with $\boldsymbol{\xi}$ and integrating over the coronal volume generates the perturbed kinetic and potential energies and conservation of energy then gives

$$\frac{1}{2}\omega^2 \int \rho\xi^2 \, dV = \delta W, \tag{7.18}$$

where

$$\delta W = -\frac{1}{2}\int \boldsymbol{\xi}\cdot F(\boldsymbol{\xi})\, dV. \tag{7.19}$$

CHAPTER 7: STABILITY AND ERUPTION OF PROMINENCES

Obviously, if the main interest is to show that the plasma is unstable, then it is only necessary to show that δW is negative. In general, if it is possible to prove that

$$\delta W > 0,$$

for all allowable ξ satisfying the boundary conditions, then the equilibrium is *stable*. However, if there is a ξ such that

$$\delta W < 0,$$

then the plasma is definitely *unstable* (Laval et al, 1965).

δW can be rewritten as

$$\delta W = \frac{1}{2} \int \left\{ \frac{1}{\mu} |\nabla \times (\xi \times B_0)|^2 + \frac{\nabla \times B_0}{\mu} \cdot [\xi \times (\nabla \times (\xi \times B_0))] + (\xi \cdot \nabla p_0) \nabla \cdot \xi \right. $$
$$\left. + \gamma p_0 (\nabla \cdot \xi)^2 + (\xi \cdot g) \nabla \cdot (\rho_0 \xi) \right\} dV, \qquad (7.20)$$

where the surface integrals are assumed to vanish due to the photospheric boundary conditions. A detailed derivation of (7.20) is given in Roberts (1967).

Newcomb (1960) used (7.20) to study the stability of *cylindrical* equilibria, of the type described in Section 7.2, when gravity is negligible, by selecting the Fourier modes

$$\xi = (\xi_r \cos(m\theta + kz), \xi_\theta \sin(m\theta + kz), \xi_z \sin(m\theta + kz)). \qquad (7.21)$$

After some tedious algebra, and the use of Calculus of Variations, (see Newcomb, 1960 for details), the stability of the magnetic field is determined by the zeros of the solution to the differential equation

$$\frac{d}{dr}\left(f \frac{d\xi_r}{dr}\right) = g\xi_r, \qquad (7.22)$$

where the functions f and g are given by

$$f = \frac{r(mB_\theta + krB_z)^2}{m^2 + k^2 r^2},$$

and

$$g = \frac{2\mu k^2 r^2}{m^2 + k^2 r^2} \frac{dp}{dr} + \frac{m^2 + k^2 r^2 - 1}{r(m^2 + k^2 r^2)}(mB_\theta + krB_z)^2 + \frac{2k^2 r}{(m^2 + k^2 r^2)^2}(k^2 r^2 B_z^2 - m^2 B_\theta^2).$$

To illustrate the method, consider the situation where f does not vanish and the boundary conditions are

$$\xi_r(r = a) = \xi_r(r = b) = 0;$$

then starting the integration of (7.22) at $r = a$, the three possible cases are shown in Figure 7.3. Curve 1 corresponds to the solution when the magnetic field is stable, curve 2 occurs at marginal

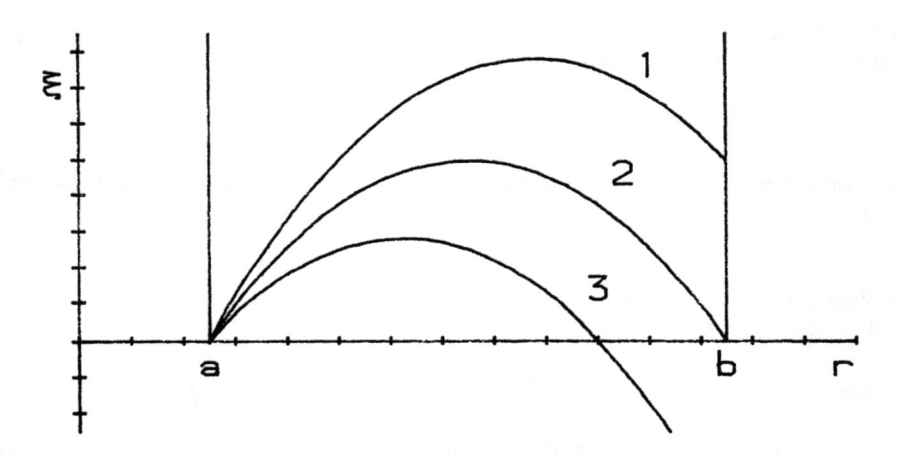

Figure 7.3 Three possible solutions to the Euler-Lagrange equation (7.22)

stability and curve 3 when the equilibrium is unstable. Using δW, it is easy to prove instability (see Wesson, 1981, for details) for curve 3.

To study the stability of more general equilibria, trial functions must be used that are more general than those chosen in equation (7.21). In particular, the correct boundary conditions must be included in the trial functions.

7.3.3 NON EQUILIBRIUM

An alternative way to study the stability and, in particular, the eruption of a prominence is to investigate conditions for which no equilibrium state exists. Here the suggestion is that the prominence magnetic field evolves slowly through a sequence of, presumably stable, equilibria until a catastrophe point is reached beyond which no neighbouring solution exists. Then, since the various forces are no longer in balance, the field must evolve dynamically on an Alfvénic time-scale.

To illustrate the basic ideas, the approach of Zwingmann (1987) is followed. Variations along the length of the prominence are neglected and so the equilibria are assumed to depend on only two spatial coordinates, for example x and z. The basic state satisfies the equation of motion (1.2) in the form

$$\nabla p = \mathbf{j} \times \mathbf{B} + \rho \mathbf{g}, \qquad (7.23)$$

where the magnetic field may be prescribed in terms of a flux function A, the y-component of the magnetic vector potential, as

$$\mathbf{B} = (-\frac{\partial A}{\partial z}, B_y(A), \frac{\partial A}{\partial x}).$$

Furthermore, if the temperature is taken initially as uniform, equation (7.23) may be written as

$$-\nabla^2 A = \frac{\partial}{\partial A}\left(\mu p(A) e^{-z/H} + \frac{1}{2} B_y^2(A)\right). \qquad (7.24)$$

This equation has been derived by several authors (see Low (1982) and references therein). However, while the unknown p(A) may be obtained from knowledge of the photospheric pressure and

CHAPTER 7: STABILITY AND ERUPTION OF PROMINENCES   175

the line of sight magnetic field, $\partial A/\partial x$, $B_y(A)$ should be obtained from the footpoint displacement. The footpoint displacement is given by

$$\Delta y(x) = \int_{x_1}^{x_2} \left(B_y/B_x\right) dx, \tag{7.25}$$

where the integrand is evaluated along a field line. This converts equation (7.24) into an integro-differential equation. However, Zwingmann (1987) rewrote the field in terms of Euler potentials. In a similar way, it is possible to solve the above equations if B_y is written in terms of suitable functions that allow the integral in equation (7.25) to be readily performed. Defining,

$$B_y = \frac{\partial \beta}{\partial z}\frac{\partial A}{\partial x} - \frac{\partial \beta}{\partial x}\frac{\partial A}{\partial z}, \tag{7.26}$$

equation (7.25) becomes

$$\Delta y(x) = \beta(x_2, 0) - \beta(x_1, 0). \tag{7.27}$$

Equations (7.24), (7.26) and (7.27) now define the equilibrium and are equivalent to those solved by Zwingmann (1987). Zwingmann selected photospheric shear and pressure profiles that can be adjusted by the parameters λ_s and λ_p, respectively. It is then possible to plot the energy of this system as these parameters are varied. Typical results are shown in Figure 7.4 and the resulting control surface, which indicates the region of multiple solutions, is displayed in Figure 7.5.

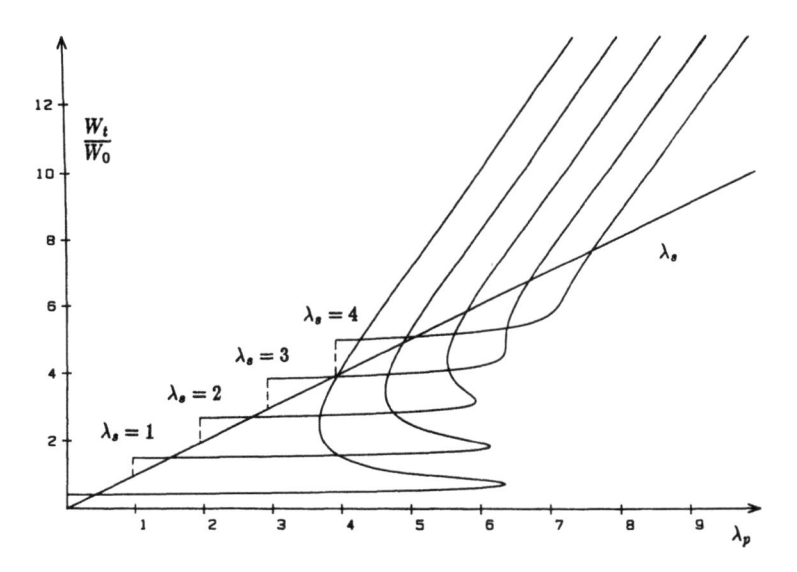

Figure 7.4 An existence diagram of the equilibrium problem in the $\lambda_s - \lambda_p$ plane (from Zwingmann 1987)

The conclusion from Zwingmann's calculation is that a catastrophe point exists when the pressure is increased slowly *only when the footpoint displacement is sufficiently small*. Increasing the shearing motion causes a reduction of the critical value of the pressure but the catastrophe point actually disappears when the shear parameter, λ_s, exceeds a typical value of 3. The exact value

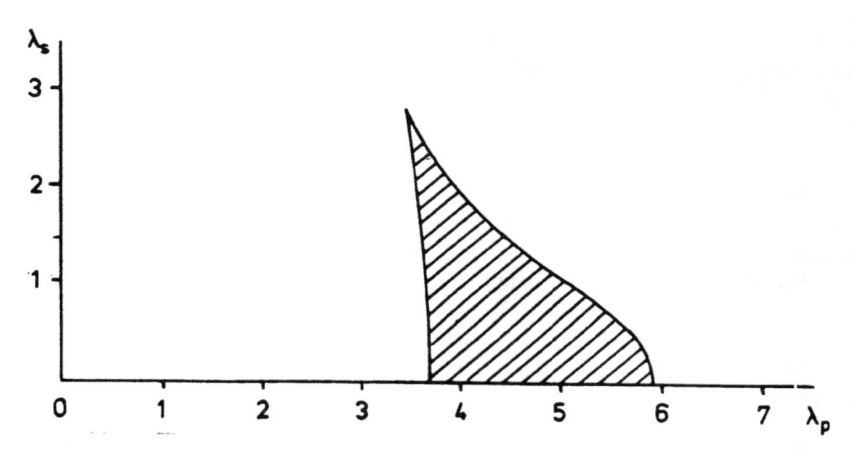

Figure 7.5 Control surface to Figure 4. Multiple solutions exist only in the shaded region (from Zwingmann 1987)

depends on the particular equilibrium studied. Thus, shearing a force-free arcade *does not* produce a catastrophe point which suggests that a prominence existing in such an arcade is unlikely to erupt.

However, the analysis of Zwingmann (1987) allows the stability of 2D equilibria to be investigated only against 2D disturbances. He shows that the solutions lying on the lower branch are stable to these coronal displacements. Now the question is how the stability is changed when fully 3D perturbations are considered. One possibility is that the fields are unstable to 3D disturbances before the catastrophe point is reached. If this happens, the field and prominence may either erupt or, if the instability saturates at a low level, a fully 3D equilibrium may exist. In this latter case, increasing the parameters may or may not lead to a catastrophe point. Such a situation has still to be investigated.

7.4 Effect of the Dense Photosphere

7.4.1 PHYSICAL ARGUMENTS

The first detailed stability calculation of coronal magnetic fields was due to Anzer (1968). By using the energy method of Bernstein *et al* (1958), he was able to prove that all force-free fields are unstable to kink instabilities. This somewhat surprising result is counter to intuition since prominence magnetic fields are observed to exist for much longer than an Alfvén travel time. In addition, the energy required to drive solar flares is thought to be stored in the coronal field and the prominence support mechanism is due to the coronal magnetic field so that the field *cannot* be in a potential state. Raadu (1972) realised that the coronal fields were not infinite in length, as assumed by Anzer, but instead entered the photosphere after a finite distance. Now the density of the photosphere greatly exceeds the coronal value and, because of the thinness of the transition region and chromosphere, coronal disturbances will "see" a sudden increase in density at the footpoints of the fieldlines. Raadu studied the stability of force-free fields and modelled the dense photosphere by assuming that the coronal displacement, ξ, vanishes at the photospheric boundary. Then he used a trial function in the energy method to show that a coronal loop could become unstable if it is sufficiently twisted. This analysis does not actually show that such *photospheric line tying* stabilises the equilibrium since he used a trial function.

Notice that the component of $\boldsymbol{\xi}$ parallel to the equilibrium field does not enter the above calculation when the fields are force-free but only when pressure gradients are included. It is the choice of boundary conditions for $\boldsymbol{\xi} \cdot \mathbf{B}$ in modelling line-tying that has been the subject of debate. Hood and Priest (1979), following Raadu, assumed the same form of trial function and so chose all components of $\boldsymbol{\xi}$ to vanish. These conditions are commonly known as the *rigid plate conditions*. Some of the other authors to use these conditions are Low (1977), Hood and Priest (1980,1981), Schindler *et al* (1983). In this case the photosphere is energetically decoupled from the corona.

On the other hand, Foukal (1976) reported observations of large downflows at the base of coronal loops and this was simulated by several authors (Van Hoven *et al*, 1980; Einaudi and Van Hoven 1981,1983; Migliuolo *et al* 1984; Hood 1983, 1984) by relaxing $\xi_{\|} = 0$ to allow a flow along the magnetic field at the photosphere. The perpendicular components are still taken as zero, in accordance with Raadu. These conditions are referred to as *flow through conditions*. To obtain the right number of boundary conditions, two conditions on $\xi_{\|}$ must be specified. The choice of conditions is usually restricted to those that energetically isolate the corona, otherwise the energy method cannot be used easily (see Einaudi and Van Hoven, 1981).

A third set of photospheric boundary conditions was used by An (1982,1984), who argued that not only should $\boldsymbol{\xi} = 0$ but also $\mathbf{B}_1 = 0$ should hold, where $\mathbf{B}_1$ is the perturbed magnetic field. However, this overspecifies the problem when studying ideal instabilities since it gives 12 boundary conditions for a sixth-order system. This can only result in a trivial solution to the linearised equation of motion but this is obscured in his work since he uses a trial function in the energy method. However, when resistivity is included, the order of the equations increases and his choice of conditions may indeed be applicable. The main effect of resistivity, in this case, will be to introduce a resistive boundary layer through which the ideal conditions are modified into the resistive ones.

Three choices of photospheric conditions have been given and, apart from the last set, plausible arguments can be made for either set. What does the sun actually do? A simple physical argument would suggest that the photosphere is unable to respond to changes in the corona, since the coronal time-scale is substantially shorter than the photospheric time-scale. This would suggest that the rigid plate conditions are the correct choice (Rosner *et al*, 1984). In a similar manner, consider an incident sound wave train impinging on the photosphere. Then, as the wave approaches the transition region the density starts to rise, while the pressure remains approximately constant. The sound speed starts to drop. A simple calculation shows that over 99% of the wave amplitude is reflected at the interface and less than 1% is transmitted. This can be simulated by setting $\xi_{\|} = 0$, the rigid plate conditions. However, there is the possibility that these simple physical ideas are just too simple. A partial answer can be obtained by including a density variation along a field line and using the *ballooning approximation* or a localised analysis as follows.

7.4.2 BALLOONING MODES

Following the approach of Connor *et al* (1979) and Dewar and Glasser (1983), a WKB solution of the form

$$\boldsymbol{\xi} = \boldsymbol{\xi} exp(inS(\mathbf{r})), \tag{7.28}$$

is sought, where n, the ratio of the equilibrium length-scale to the perturbation length-scale, is assumed large. Defining the wave vector as $\mathbf{k} = \nabla S$, then the stable waves are eliminated if

$$\mathbf{k} \cdot \mathbf{B} = 0. \tag{7.29}$$

Substituting into the linearised equation of motion (7.8) gives two coupled equations for the Alfvén and slow mode wave amplitudes. Hood (1986b) assumed a density profile that increased sharply through a small region and by solving the ballooning equations in each region was able to show that the rigid plate conditions are indeed the most relevant photospheric boundary conditions.

7.5 Coronal Arcades

7.5.1 DISTRIBUTED CURRENT MODELS - ERUPTIVE INSTABILITY

Before investigating the stability of magnetic arcades, it is illuminating to rearrange the potential energy in terms of physically meaningful quantities. Following Bateman (1980),

$$\delta W = \frac{1}{2}\int d^3x \Big\{ \frac{1}{\mu}\mid \mathbf{B}_{1\perp}\mid^2 + \mu\mid \frac{1}{\mu}\mathbf{B}_{1\parallel} - \mathbf{B}_0(\boldsymbol{\xi}\cdot\nabla p_0)/B_0^2\mid^2 + \gamma p_0\mid \nabla\cdot\boldsymbol{\xi}\mid^2 \\ + \frac{\mathbf{j}_0\cdot\mathbf{B}_0}{B_0^2}(\mathbf{B}_0\times\boldsymbol{\xi})\cdot\mathbf{B}_1 - 2(\boldsymbol{\xi}\cdot\nabla p_0)(\boldsymbol{\xi}\cdot\boldsymbol{\kappa}) + (\mathbf{g}\cdot\boldsymbol{\xi})\nabla\cdot(\rho_0\boldsymbol{\xi})\Big\}, \qquad (7.30)$$

where $\mathbf{B}_1 = \nabla\times(\boldsymbol{\xi}\times\mathbf{B}_0)$ and $\boldsymbol{\kappa} = \mathbf{b}_0\cdot\nabla\mathbf{b}_0$ is the normal curvature in terms of $\mathbf{b}_0 = \mathbf{B}_0/B_0$. The first three terms represent the potential energy of the stable Alfvén, fast magnetoacoustic and acoustic waves. The last three terms are, respectively, the driving terms behind the current-driven kink modes, the pressure-driven localised interchange modes and the gravity-driven Rayleigh-Taylor modes. As already mentioned, the main stabilising effect is due to the line-tying effect of the dense photosphere (Raadu, 1972). To analyse the stability of two-dimensional coronal arcades, with one ignorable coordinate, Eq (7.30) must be rearranged further (see Schindler *et al* 1983, Hood 1984, Melville *et al* 1986, and Zweibel 1981). Minimising δW results in complicated partial differential equations and as such it is difficult to obtain general results. Instead, the stability of various equilibria is analysed individually.

The results of Hood and Priest (1980a), Hood (1983), Zweibel (1981), Melville *et al* (1987), Schindler *et al* (1983), Cargill *et al* (1986), Einaudi and Van Hoven (1981) suggest that photospheric line-tying strongly stabilises coronal arcades since the kink mode is prohibited. This can be seen by considering Fourier modes in the azimuthal direction. Since line-tying is applied at $\theta = 0$ and $\theta = \pi$, the azimuthal wave number must be restricted to even integers and so the $m = 1$ kink mode is not allowed. In addition, gravity can either stabilise or destabilise the plasma depending on the density distribution. Cargill *et al* (1986) used a finite difference scheme in the azimuthal direction, to derive a system of differential equations for the radial variation. This system was then finite-differenced in the radial variable and the simple stability criterion, in terms of the zeros of ξ_r, as discussed in section 7.3.2, is replaced by a similar criterion based on the zeros of the determinant of the coefficients of $\boldsymbol{\xi}$. Einaudi and Van Hoven (1981) used instead a truncated Fourier series to study the stability of coronal loops and the two methods give good agreement for critical coronal conditions. These approaches correctly predict the behaviour of global modes but cannot easily pick up highly localised modes.

Two main ways of making an arcade go unstable and erupt have been proposed. The first (Hood and Priest, 1980a) is that a magnetic island (see Figure 7.6) in a force-free field may produce an instability. If the island, really helical in nature, exceeds a critical height above the photosphere then an active-region filament can be destabilised since a significant fraction of the field no longer benefits from the line-tying stabilisation.

The second suggestion is by Cargill *et al* (1986) and Hood and Anzer (1987) namely that substantial pressure gradients, due to the presence of an active-region filament, create an instability, even

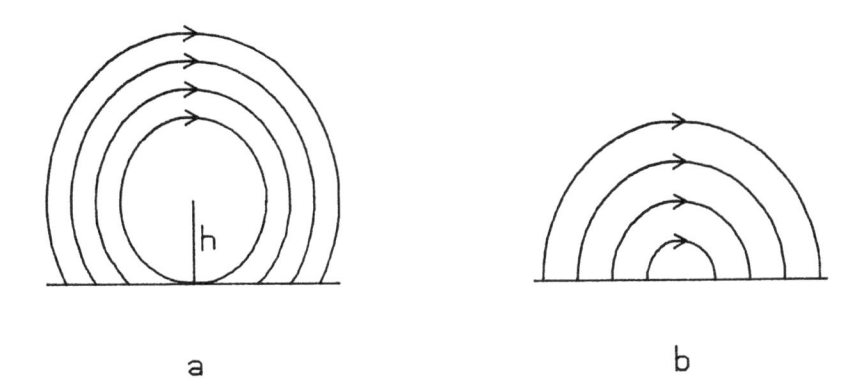

CHAPTER 7: STABILITY AND ERUPTION OF PROMINENCES   179

Figure 7.6. Two possible projections of an active region arcade. In (a), h is the height of the island above the photosphere.

in the absence of magnetic islands. Indirect support for this conjecture comes from Zwingmann (1987) on the basis of equilibrium models, as discussed in Section 7.3.3.

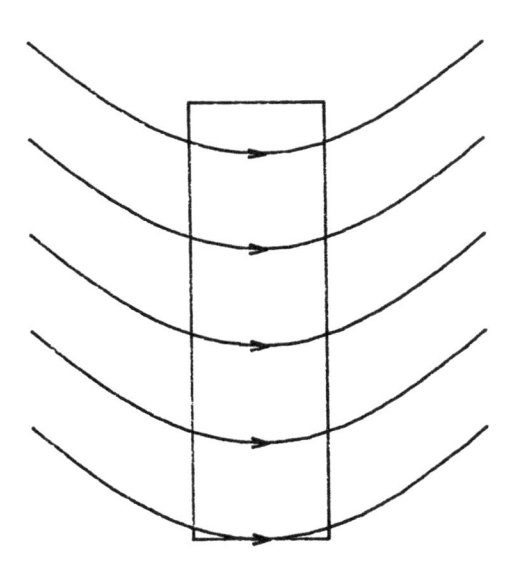

Figure 7.7. Schematic representation of the Kippenhahn-Schlüter prominence field.

Furthermore the Kippenhahn-Schlüter (1957) prominence model, sketched in Figure 7.7 has been investigated by several authors. Migliuolo (1982) and Zweibel (1982) showed that a shearless field was stable to arbitrary displacements and a completely general stability proof for a sheared field of this type was presented by Galindo-Trejo and Schindler (1984). However, this type of analysis only deals with the stability of the internal structure of the prominence and no account has been taken of the external field that eventually links down to the photosphere.

7.5.2 LOCALISED MODES - SMALL SCALE STRUCTURE

Localised or ballooning modes are driven by a gas pressure gradient and result in an interchange instability. They are characterised by small wavelengths and large variations across magnetic field lines but have a long length-scale along the field. The slowness of the variations along the field reduces the stabilising effect of waves generated by magnetic tension.

The theory of ballooning instabilities was developed by Connor et al (1979) and, with a rigorous treatment of a WKB approach, by Dewar and Glasser (1983). These modes are interesting in the solar context since they allow photospheric line-tying to be included in a simple manner. Hood (1986a) used the approach of Dewar and Glasser to investigate the effect of line-tying on ballooning modes and to derive a simple test for instability. The displacements are assumed to have the form taken in equation (7.28). Substitution of ξ into equation (7.20) results in a simplified form of the potential energy. This allows a simple condition for instability to be derived (since a trial function has been used) that is similar to Suydam's criterion (Suydam 1958). Hood (1986a) and De Bruyne and Hood (1988) obtained, for cylindrically symmetric arcades with the axis of symmetry on the photosphere,

$$a\mu\frac{2}{r}\frac{dp}{dr} + bB_z^2\left(\frac{dq/dr}{q}\right)^2 + c\frac{B^2}{r^2} + d\frac{4\gamma\mu p B_\theta^2}{r^2(\gamma\mu p + B^2)} < 0, \tag{7.31}$$

where $q = rB_z/B_\theta$ and the constants a, b, c and d are given by

$$a = 2x - 3\sin x + x\cos x,$$
$$b = \frac{x^3}{12} - \frac{1}{4}((x^2 - 2)\sin x + 2x\cos x)),$$
$$c = \frac{x^2}{\pi^2}(x - \sin x),$$
$$d = \frac{1}{x}((x^2 - 4)\cos x - 4x\sin x + x^2 + 4),$$

and x is chosen to make the left-hand side of (7.31) as small as possible. A good choice is given by

$$x^2 = -2\mu\pi^2 rp'/B^2.$$

The first term in (7.31) drives the instability through an adverse pressure gradient whereas the second term stabilises due to shear in the magnetic field. These first two terms are essentially Suydam's criterion. The last two terms result from line tying and correspond to the stabilising effects of tension and compression due to standing Alfvén and slow mode waves.

Zweibel (1981) and Melville et al (1986, 1987) analysed the stability of isothermal, sheerless arcades including line-tying and gravitational effects. Localised modes, are present that are centred about a particular magnetic flux surface. Melville et al (1987) studied the field

$$\mathbf{B} = \left(\frac{\partial A}{\partial y}, -\frac{\partial A}{\partial x}, 0\right),$$

with the flux function A given by the either of the double solutions,

$$A_\pm = \log(x^2 + (y - h_\pm)^2 + \beta/4) - \frac{1}{2}\alpha y,$$

CHAPTER 7: STABILITY AND ERUPTION OF PROMINENCES 181

where
$$h_\pm = \pm(1 - \beta/4)^{1/2},$$

β is the plasma beta at the origin and α is the ratio of the horizontal length-scale to the pressure scale height. The importance of gravity is increased as the parameter α increases. They find that, as α increases the arcade is destabilised but that the unstable region becomes more localised. Since these instabilities are driven by pressure gradients, it is unlikely that they will completely destabilise the coronal and prominence magnetic fields since the internal energy of the plasma is generally much smaller than the magnetic energy in the corona. However, they automatically provide a mechanism for generating small length scales and so they provide one possible explanation for the observed small-scale structure in prominences. These small length-scales could also result in enhanced transport coefficients, allowing the plasma to diffuse across the field, as well as contributing towards the unknown coronal heating function.

7.5.3 ARCADES CONTAINING A CURRENT SHEET

Kippenhahn and Schlüter (1957) modelled a prominence by an infinitely thin current sheet surrounded by a potential coronal field. This approach has been extended by Kuperus and Raadu (1974), Malherbe and Priest (1983) and Anzer and Priest (1985). The most complete stability analysis of current sheet models is due to Anzer (1969). By manipulating δW inside the current sheet the potential energy reduces to a simple form since the coronal volume integral and the intermediate surface integral do not contribute to leading order.

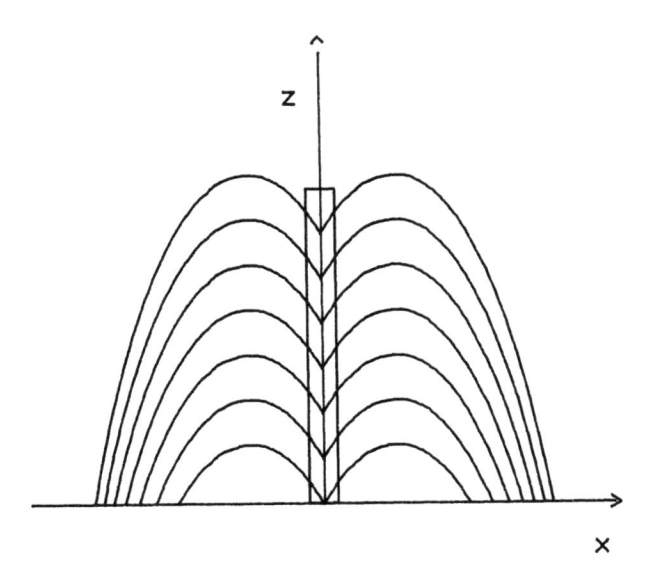

Figure 7.8. A current sheet prominence model situated on the z-axis.

In terms of the cartesian coordinate system, as shown in Figure 7.8, the equilibrium is given by

$$g\rho(z) = B_z(x=0,z)[B_x(z)]/\mu,$$

$$\lim_{\delta \to 0} \{B_z(x = \delta, z) + B_z(x = -\delta, z)\} = 0$$

and

$$p(x = 0, z) = \frac{1}{8\mu}[B_z]^2,$$

where $\rho(z)$ is the current sheet mass density, $-[B_z(z)]/\mu$ is the sheet current and where

$$[B_z] = \lim_{\delta \to 0} \{B_z(x = +\delta, z) - B_z(x = -\delta, z)\}$$

is the jump in B_z across the prominence. The results that for every function $f(x, y, z)$, with $\partial f/\partial x$ finite,

$$\lim_{\delta \to 0} \int_{-\delta}^{+\delta} f \frac{\partial B_z}{\partial x} dx = f(x = 0)[B_z]$$

and

$$\lim_{\delta \to 0} \int_{-\delta}^{+\delta} f B_z \frac{\partial B_z}{\partial x} dx = 0,$$

may be used to manipulate the perturbed potential energy, δW, into the form

$$\delta W = \frac{1}{2} \int \int \left\{ \frac{dB_x(x=0)}{dz}[B_z]\xi_x^2 - B_x(x=0)\frac{d[B_z]}{dz}\xi_z^2 \right\} dy dz.$$

Thus, sufficient conditions for stability become

$$[B_z]\frac{dB_x}{dz}(x = 0) \geq 0 \qquad (7.32)$$

$$B_x(x = 0)\frac{d[B_z]}{dz} \leq 0, \qquad (7.33)$$

at all heights z. Anzer (1969) shows that these conditions are in fact necessary and sufficient. Conditions (7.32) and (7.33) represent restrictions on the initial coronal field for the formation of a stable prominence. If at some time either of (7.32) and (7.33) are violated an instability sets in. Anzer (1969) interprets the violation of condition (7.33) in terms of a modified Rayleigh-Taylor instability. This analysis was extended by Wu (1987) to include curved current sheet models and he showed that the curved static sheet models of Wu and Low (1987) are all unstable.

The prominence has been represented by a line current in the work of Van Tend and Kuperus (1978). The equilibrium is represented by a balance between (i) a Lorentz force between the current and the background magnetic field, (ii) a Lorentz force representing the interaction of the current and a mirror current in the photosphere and (iii) the gravitational force. Thus,

$$\mu I^2/h = IB + mg,$$

where I is the current in the prominence per unit length, B is the background coronal field, m is the mass of the prominence per unit length and h is its height above the photosphere. The solution of this equation gives the current in terms of the height of the prominence, or visa-versa. Because of the quadratic nature of the equation and the particular choice for the background horizontal magnetic field, there exist catastrophe points. If the prominence current is increased at the lower point, there is a sudden jump in the height of the prominence and Van Tend and Kuperus associate this with an eruption of the prominence. A simple stability analysis (Van Tend and Kuperus, 1978; Kuperus and Van Tend, 1981) suggests that as the current in the prominence slowly increases it rises through a series of stable equilibria until the catastrophe point is reached and the forces no longer balance. Then the filament erupts. Recently this model has been extended to more realistic coronal situations since the prominence current and the coronal magnetic field are not independent but are linked in a non-linear manner (Amari and Aly, 1988; Demoulin and Priest, 1988). MHD modes, both compressional and kink, have been neglected and these may have a lower threshold for instability than the purely vertical displacements have been considered.

7.6 Thermal Stability

Most prominence stability investigations have just dealt with the magnetic field and assumed the plasma to be isothermal. Obviously this is a major simplification and the main attempts to understand thermal instability have been restricted to the condensation from a hot isothermal plasma and not to a typical prominence structure. However, it is still important to discuss thermal stability in general terms.

The general properties of radiative instabilities were discussed by Parker (1953), Field (1965) and by Heyvaerts (1974) for a uniform medium. Thermal instabilities depend on the shape of the radiative loss function. For example, the equilibrium is given by

$$\mathbf{B} \cdot \nabla\left(\frac{\kappa_\parallel}{B^2}\mathbf{B} \cdot \nabla T\right) = L = p^2 \chi T^{\alpha-2} - h(\rho, T), \tag{7.34}$$

where the optically thin radiative loss function χT^α is a piecewise continuous function. The qualitative form of the function L is sketched in Figure 7.9, assuming constant pressure and uniform heating per unit volume.

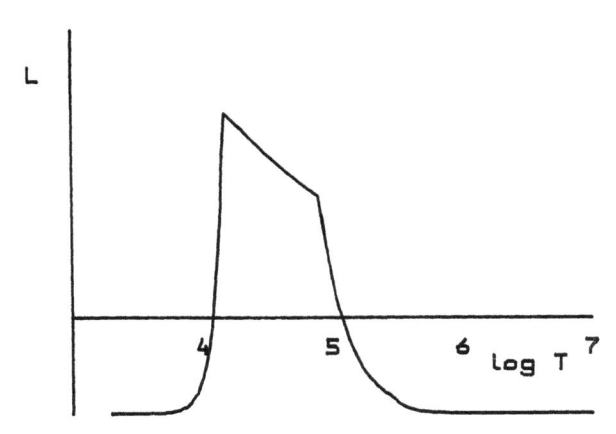

Figure 7.9. The total loss function, L, in Equation (7.34)

Linearising about a uniform isothermal basic state and assuming isobaric disturbances so that $\delta p = 0$, the energy equation (1.5) can be written as

$$\frac{\gamma p}{\gamma - 1} \frac{\partial \delta T}{\partial t} = -\frac{(\mathbf{k} \cdot \mathbf{B})^2}{B^2} \kappa_\parallel \delta T - \frac{\partial L}{\partial T} \delta T, \tag{7.35}$$

where δT is the temperature perturbation and $\mathbf{k}$ is the wave vector. When thermal conduction is negligible ($\mathbf{k} \cdot \mathbf{B} = 0$), the plasma is unstable if

$$\frac{\partial L}{\partial T} < 0, \tag{7.36}$$

or in terms of α, if $\alpha < 2$. Thermal conduction helps to stabilise the radiative instability. This instability is linear and so the temperature can either increase (resulting in heating) or decrease (resulting in cooling). Hence, this may be important in the formation stage of a prominence's life since localised cooling will trigger the thermal instability and result in enhanced radiation and more cooling. Once prominence temperatures are reached, $\partial L/\partial T > 0$, so stabilising the radiative mode. Thermal instability has been investigated in coronal loops, i.e. a rigid tube, by Habbal and Rosner (1980), Antiochos (1979) and Hood and Priest (1980b) and with more detailed attention paid to the chromospheric boundary conditions by McClymont and Craig (1985). Although this type of analysis considers more realistic plasmas, the basic temperature profile is coronal in nature and not applicable to a prominence.

Attempts have been made to couple the radiative mode to MHD instabilities, through non-isobaric disturbances but attention has so far been focussed on an infinite plasma without chromospheric/photospheric boundary conditions. The main workers in this area so far are An (1987), Steinolfson and Van Hoven (1984), Mok and Van Hoven (1987), Sparks and Van Hoven (1985). In general, they find that coupling the radiative mode to the tearing mode causes an enhanced growth rate in the latter mode. However, much more work is still required on this research topic.

So the question still remains, how does a prominence disappear thermally? (Wagner, 1987; Mouradian et al, 1987). To evaporate a prominence, the prominence temperature must rise above the temperature corresponding to the maximum in the total loss function L so that the thermal instability can operate in the other direction. Then the thermal runaway of Kahler and Kreplin (1970) can take place. This temperature is in general less than 10^5 K. Malherbe and Forbes (1987) have suggested that the appearance of a hot region in the neighbourhood of a prominence could enhance conduction into the prominence and thereby heat it. This simple order of magnitude model must be investigated in more detail to check on its predictive properties.

Further references to thermal stability and their applications to the formation of prominences can be found in Chapter 5.

7.7 Resistive Instabilities - Tearing Modes

7.7.1 INTRODUCTION

The inclusion of uniform resistivity in the induction equation,

$$\frac{\partial \mathbf{B}}{\partial t} = \nabla \times (\mathbf{v} \times \mathbf{B}) + \eta \nabla^2 \mathbf{B}, \tag{7.37}$$

allows a new class of instabilities. Three types of resistive instability are discussed in Furth, Killeen and Rosenbluth (1963) (FKR) namely (i) the resistive gravity mode, (ii) the rippling mode and (iii) the tearing mode, which are driven by (i) gravity and an unfavourable density distribution or by curvature, (ii) gradients in the resistivity (not considered here) and (iii) a non-uniform current density, respectively.

In this section, only the tearing mode is studied and the development follows Bateman (1980) and Priest (1985). First of all, an intuitive description of the tearing mode and its nonlinear development to steady magnetic reconnection is presented. In the ideal limit, the field is frozen into the plasma and field lines are not allowed to break and reconnect (field lines retain their identity). Thus, the effect of pressing together oppositely directed field lines of a current sheet is just to build up the magnetic field strength at the centre. However, when η is included, the field may now diffuse relative to the fluid (as in Fig 7.10(c)) and reconnect to change its topology (as in Fig 7.10(d)). Thus, lower energy states are now accessible and new instabilities can take place that were absent in the ideal limit. Two qualitative descriptions are as follows.

CHAPTER 7: STABILITY AND ERUPTION OF PROMINENCES   185

The time-scale for diffusion for a field that changes rapidly over a length l (the width of the current sheet) is l^2/η. Thus, when the field is squeezed together, as in Fig 7.10(b), l becomes smaller and so the field diffuses faster at the point of squeezing than elsewhere. This leads to the situation in Fig 7.10(c). Hence, the field lines must reconnect. The restoring force due to magnetic tension is reduced if the wavelength of the disturbance is long.

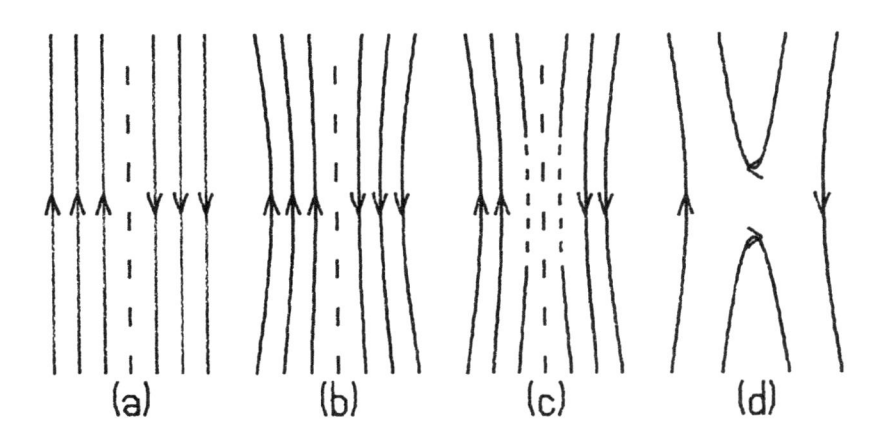

Figure 7.10 Magnetic Reconnection

Alternatively, the current sheet may be represented by equally spaced current-carrying wires. A small displacement that causes some wires to bunch together will then continue because of the mutual attraction of the wires, as shown in Figure 7.11.

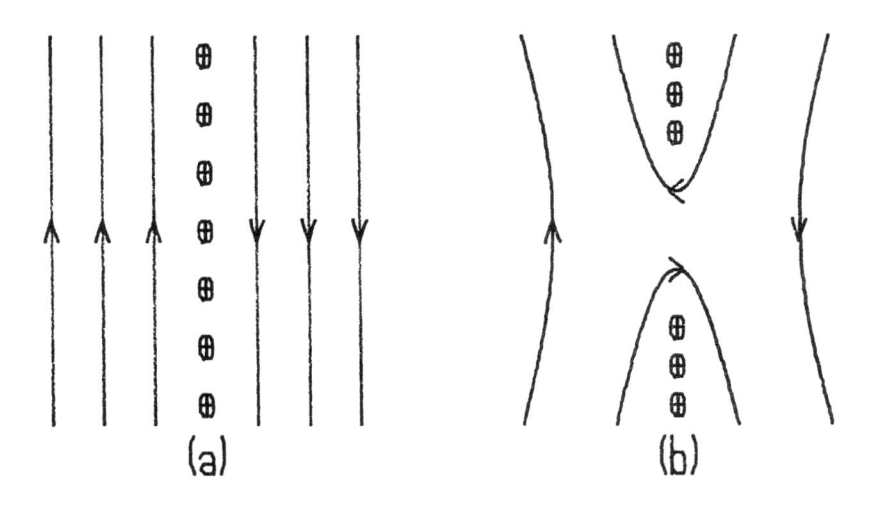

Figure 7.11 The tearing mode - an attraction of line currents

7.7.2 ESTIMATE OF THE TEARING MODE GROWTH RATE

Considering the sheared cartesian equilibrium magnetic field

$$\mathbf{B}_0 = B_{0y}(x)\hat{\mathbf{y}} + B_{0z}(x)\hat{\mathbf{z}}, \quad (7.38)$$

the linearised equations for an *incompressible* plasma are (7.3), (7.5) and $\nabla \cdot \mathbf{v} = 0$. Taking Fourier components of the perturbed variables as

$$f(x)e^{i(k_y y + k_z z) + st},$$

and using the x- and z-components of $\nabla \times$(7.2), the x-component of (7.5) and the incompressible assumption, the resistive equations become

$$sB_{1x} = iv_{1x}(\mathbf{k} \cdot \mathbf{B}_0) + \eta\left(\frac{d^2 B_{1x}}{dx^2} - k^2 B_{1x}\right), \quad (7.39)$$

$$s\left(\frac{d^2 v_{1x}}{dx^2} - k^2 v_{1x}\right) = \frac{i(\mathbf{k} \cdot \mathbf{B}_0)}{\mu \rho_0}\left(\frac{(\mathbf{k} \cdot \mathbf{B}_0)''}{(\mathbf{k} \cdot \mathbf{B}_0)}B_{1x} + \frac{d^2 B_{1x}}{dx^2} - k^2 B_{1x}\right), \quad (7.40)$$

The incompressible assumption is valid if the instability growth rate is much smaller than the Alfvén rate and the fluid and phase velocities are subsonic. $\mathbf{k} \cdot \mathbf{B}_0$ and k^2 are defined as

$$\mathbf{k} \cdot \mathbf{B}_0 = k_y B_{0y} + k_z B_{0z},$$
$$k^2 = k_y^2 + k_z^2, \quad (7.41)$$

and dashes denote derivatives with respect to x. The first term on the right-hand side of (7.39) normally dominates the second except in the neighbourhood of a singular layer, where $\mathbf{k} \cdot \mathbf{B}_0 = 0$.

Equations (7.39) and (7.40) are nondimensionalised against a typical field strength, B_0, length-scale, a, diffusion time, $\tau_d = a^2/\eta$, and Alfvén travel time, $\tau_A = a/v_A$. (7.39) and (7.40) are rewritten as

$$\sigma B_{1x} = -v_{1x}F + (B_{1x}'' - k^2 B_{1x}), \quad (7.42)$$

$$\sigma(v_{1x}'' - k^2 v_{1x}) = S^2 F\left(-B_{1x}\frac{F''}{F} + B_{1x}'' - k^2 B_{1x}\right),. \quad (7.43)$$

All quantities are dimensionless where $F = (\mathbf{k} \cdot \mathbf{B}_0)/k$ and $\sigma = s\tau_d$.

The magnetic Reynolds number or Lundquist number $S = \tau_d/\tau_A$, the ratio of the diffusion to the Alfvén travel time, is assumed to be much greater than unity.

It is assumed that the instability will grow on a hybrid time-scale that is very much smaller than τ_d but very much greater than τ_A.

$$\tau_d \gg \tau = \tau_d/\sigma \gg \tau_A$$

or

$$1 \ll \sigma \ll S.$$

This can happen if σ depends on S as $\sigma \sim S^c$, where $0 < c < 1$. $c = 0$ is the diffusive limit and $c = 1$ is the ideal limit.

Estimates of the diffusion width (the width of the boundary layer) and the tearing mode growth rate are obtained following Bateman (1980 Chap 10). Terms like $d^2 B_{1x}/dx^2$ are only large in a thin boundary layer near the resonant surface, x_s, where $\mathbf{k} \cdot \mathbf{B} = 0$. Hence, the derivative of B_{1x}, B'_{1x}, will appear to have a discontinuity here when viewed on the large scale of the plasma. Therefore,

$$\frac{d^2 B_{1x}}{dx^2} \approx \frac{B'_{1x}(x_s + \epsilon) - B'_{1x}(x_s - \epsilon)}{2\epsilon}$$
$$= \frac{\Delta' B_{1x}}{\epsilon}, \qquad (7.44)$$

where, the standard notation is

$$\Delta' \equiv \frac{B'_{1x}(x_s + \epsilon) - B'_{1x}(x_s - \epsilon)}{2 B_{1x}}. \qquad (7.45)$$

Now, inside the resistive layer all three terms of (7.39) are of similar size and so, comparing orders of magnitude and using (7.45)

$$s = \eta \Delta' / \epsilon, \qquad (7.46)$$

and

$$i v_{1x} (\mathbf{k} \cdot \mathbf{B}_0) = \eta \Delta' B_{1x} / \epsilon. \qquad (7.47)$$

In the resistive layer $(\mathbf{k} \cdot \mathbf{B}_0)$ is approximately $(\mathbf{k} \cdot \mathbf{B}_0)' \epsilon$ and the inertia term in (7.40) becomes important. Setting

$$\frac{d^2 v_{1x}}{dx^2} = \frac{v_{1x}}{\epsilon^2},$$

and balancing with the second term on the right-hand side of (7.40) (this is the main driving term due to an increase in the sheet current), gives

$$\frac{s v_{1x}}{\epsilon^2} = \frac{i (\mathbf{k} \cdot \mathbf{B}_0)'}{\mu \rho_0} \Delta' B_{1x}. \qquad (7.48)$$

Then using (7.47) the width of the resistive layer is given in terms of the growth rate

$$\epsilon^4 = s \mu \rho_0 \eta / (\mathbf{k} \cdot \mathbf{B}_0)'^2. \qquad (7.49)$$

Substituting (7.49) into (7.46) and rearranging defines the growth rate as

$$s = \tau_d^{-3/5} \tau_A^{-2/5} \Delta'^{4/5}. \qquad (7.50)$$

s then depends on S to the power 2/5 (c = 2/5).

7.7.3 EFFECT OF LINE-TYING

Line-tying provides a strong stabilising influence on ideal MHD disturbances but its effect on resistive instabilities has not yet been investigated in detail. The standard tearing mode owes its existence to a singular surface in Fourier space. This singular layer is given by the condition that the convective term $\nabla \times (\mathbf{v} \times \mathbf{B})$ vanishes at some surface. Considering the magnetic field, (7.37), the x-component of the linearised form of (7.36) can be written as

$$s B_{1x} = s (\mathbf{B}_0 \cdot \nabla) \xi_x + \eta \nabla^2 B_{1x} \qquad (7.51)$$

The singular surface is then given by

$$(\mathbf{B}_0 \cdot \nabla) \xi_x = 0. \qquad (7.52)$$

This then implies that ξ_x is constant along a field line. However, if line-tying requires ξ_x to vanish at the photosphere, then (7.52) cannot be satisfied and so there is no longer a singular layer in the strict mathematical sense. Resistive effects become important when the two terms on the right-hand side of (7.51) are of the same order of magnitude and not just when the first term vanishes. Normally, the second term is negligible unless either the length-scales are short, as in resistive ballooning modes, or the growth rate is close to ideal marginal stability so that $s \approx 0$.

Velli and Hood (1986, 1987) investigated resistive ballooning modes in both arcade and loop geometries. They showed that there is always a resistive instability but that its nature changes from pure diffusion, $\tau \approx S^{-1}$, when the equilibrium is ideally stable through a fractional power, $\tau \approx S^{-1/3}$, near ideal marginal stability to the ideal growth rate when the plasma is ideally unstable. The importance of these instabilities is not so much in predicting the global stability of prominences but more due to the fact that they automatically generate small scale structures and produce enhanced transport coefficients.

Line-tied tearing modes have only so far been investigated when a singular surface is still present. This occurs when, for example, $B_z(x) = 0$ at $x = 0$, say, and $\partial/\partial y = 0$. However, this reversal in the line of sight magnetic field is not really applicable to prominence fields.

Mok and Van Hoven (1982) studied the tearing mode in cylindrical loops with the azimuthal wavenumber $m = 0$. In addition, they assumed that the loop is very long so that there is no coupling between Fourier modes. Velli and Hood (1988) relaxed this assumption and investigated tearing modes in a cartesian geometry. A simple discretisation of the wavevector, $k = 2n\pi/L$, and the standard tearing mode analysis is applicable for the higher harmonics but the fundamental mode is strongly stabilised by line-tying. The usual $S^{-3/5}$ scaling of the growth rate is obtained. The internal structure of the resistive layer is also modified by the coupling of modes that have different singular layer widths.

7.8 Simple Model of Prominence Eruption and a Coronal Mass Ejection

Coronal mass ejections occur when large amounts of coronal plasma are ejected from the Sun. In most cases, they are associated with the eruption of an underlying prominence and, in the case of active-region filaments, with a two-ribbon flare. The structure of a coronal mass ejection includes a bubble moving out from the Sun at velocities ranging from $100\,kms^{-1}$ to over $1000\,kms^{-1}$, followed by a less dense cavity, which is assumed to contain a stronger magnetic field, and with an erupting prominence underneath.

Pneuman (1980) represented the coronal mass ejection by a curved flux tube or a bubble and derived simplified equations of motion to describe their overall motion. This was extended by Anzer and Pneuman (1982) to include a prominence beneath the bubble driven by reconnection and by Steele and Priest (1988), who brought together the previous ideas in a more realistic analysis in which the whole structure reaches a state of nonequilibrium and erupts with reconnection below the prominence driven by the eruption. Following Steele and Priest, order of magnitude equations of motion can be developed for the height, S, and width, D, of the transient and the prominence height, R, by examining the local gravitational and magnetic tension and pressure forces acting on the upper and lower surfaces of the bubble and the midpoint of the prominence.

The bubble has curvature R_a, thickness D and is at a distance S from the solar centre. The magnetic field strength is B, the plasma density is ρ and the bubble subtends an angle 2θ at the solar centre. The prominence is modelled by a twisted flux tube at a height R above the solar centre with curvature R_c, and width h. The longitudinal field along the prominence length is B_ℓ with an azimuthal field B_{az} in the same direction as that in the bubble. The filament mass density is ρ_f and the angle subtended at the solar centre is 2ϕ. The coronal cavity has a field of strength B_2 that wraps around the prominence with a fraction of this connecting down to the photosphere. The field underneath the filament and disconnected from the photosphere has strength B_3 and this increases as reconnection occurs below the prominence.

The approximate equations of motion are

$$\frac{d^2 S}{dt^2} = \frac{B_2^2}{2\mu\rho D} - \frac{B^2}{\mu\rho R_a} - \frac{GM_\odot}{S^2}, \quad (7.53)$$

$$\frac{d^2 D}{dt^2} = \frac{2 B^2}{\mu\rho D} - \frac{B_2^2}{\mu\rho D}, \quad (7.54)$$

and

$$\frac{d^2 R}{dt^2} = \frac{B_3^2 - B_2^2}{\mu h \rho_f} + \frac{B_{az}^2}{\mu\rho_f R_c} - \frac{B_\ell^2}{\mu\rho_f R_c} - \frac{GM_\odot}{R^2}. \quad (7.55)$$

In (7.53) and (7.55) the last terms on the right-hand side are due to gravity, the terms involving D or h are due to magnetic pressure, and those involving the radii of curvature are magnetic tension forces. Conservation laws of mass and magnetic flux relate ρ, B, R_a and B_2 to D, S, R and h. Further conservation laws depend on the choice of prominence geometry. Steele and Priest discuss two possibilities. Fixing all the parameters except for the magnetic flux (F) beneath the prominence, the evolution of the possible equilibria can be investigated as reconnection (below the prominence) increases F. In some circumstances, there is a catastrophe point beyond which no equilibrium solutions exist and the coronal transient begins. A typical example is shown in Figure 7.12.

This simple model couples the evolution of both the erupting prominence and the bubble (coronal transient). As the eruption continues, the velocities of the prominence and bubble tend to constant values, in agreement with observations. Obviously, 3-D instabilities cannot be studied with such a simple approach.

7.9 Conclusions and Future Work

Stability theory has been mainly applied to solar flare theory but the basic ideas can be used to study prominence stability too. However, a lot more work needs to be done once more realistic equlibrium models are available. The magnetic field in prominences is stable to ideal MHD disturbances, due to the dense photosphere which line-ties the magnetic footpoints, unless there are magnetic islands (helical field lines detached from the photosphere) that exceed either a critical height or a critical twist (Hood and Priest, 1980a). Alternatively, substantial pressure gradients must be present in order to drive ideal instability. However, ideal instabilities cannot be the final answer to the question of prominence eruption, since they do not allow a change in magnetic topology. Resistive effects must become important with the ideal instability producing a mechanism for stretching the field lines and driving flux surfaces together. Driven reconnection can rearrange the field on a time-scale that is a substantial fraction of the Alfven time. Thus, the eruption of a prominence in under an hour may be explained by such a mechanism. More details on possible causes for prominence eruption are summarised in Sections 2.4 and 7.5.1.

The fine-scale structure in prominences can owe its existence to either thermal variations (due to the thermal isolation of field lines from their neighbours) or localised instabilities that automatically generate short length-scales.

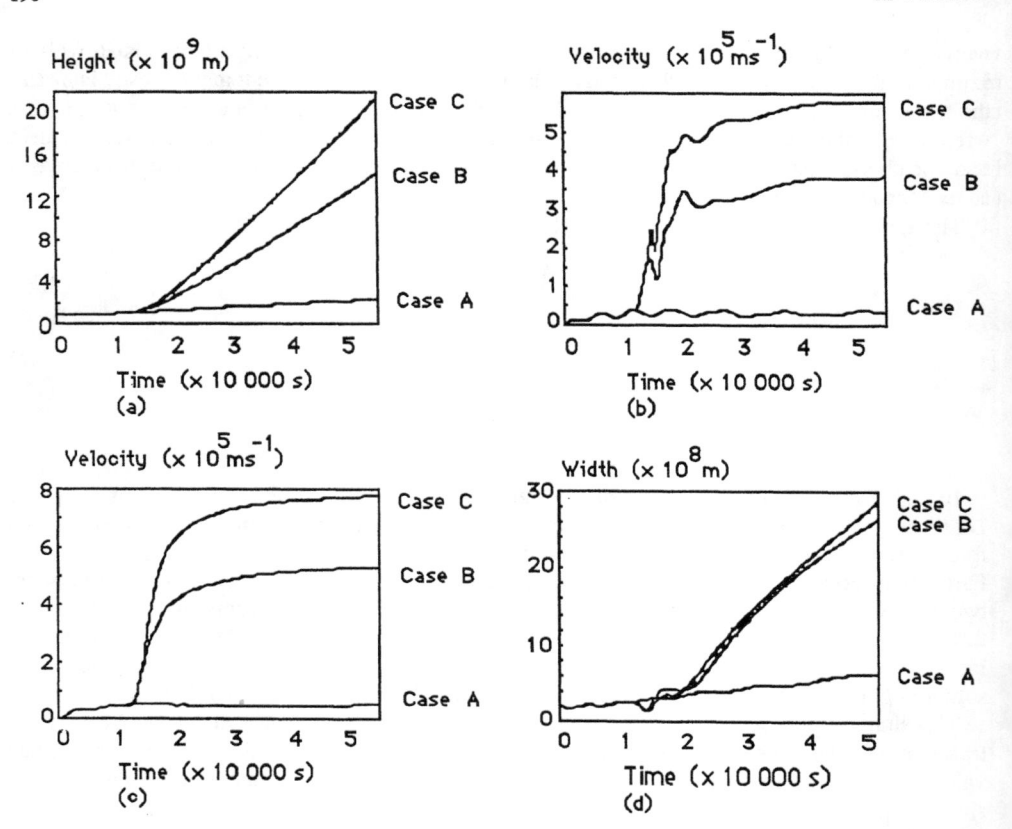

Figure 7.12 The time evolution of (a) the height of the prominence (b) the speed of rise of the prominence (c) the speed of rise of the transient and (d) the width of the transient. Cases A, B and C correspond to situations with no reconnection, and with reconnection stopping when the prominence reaches a height of 0.75 and 1 solar radius, respectively. (From Steele and Priest, 1988.)

The main areas requiring more research work are the following.

(i) Thermal effects. To date, most stability work has concentrated on purely magnetic effects that influence the overall prominence structure. All the work that couples magnetic and thermal effects has been concerned with the formation of prominences and so the temperatures have always been taken as typical coronal values. To provide a more realistic discription of the thermal disappearance of prominences, the equilibrium temperature structure must resemble a typical prominence profile.

(ii) Resistive line-tying. Line-tying has been shown to be a very strong stabilising factor in the stability of prominences to ideal MHD disturbances. With resisitivity included, it seems that line-tying does indeed influence tearing modes as well. However, the precise effect of line-tying is not yet understood and it may be that resistive instabilities are always present but their dependence on the Lundquist number, S, may be altered. For example, it is possible that the plasma is always unstable to purely diffusive modes with a very slow growth rate that is proportional to $S^{-1}\tau_A^{-1}$, where τ_A is the Alfvén timescale. When the plasma is near ideal marginal stability the growth rate may adjust to a typical tearing mode value of $S^{3/5}\tau_A^{-1}$. However, this conjecture requires detailed numerical simulation before the real effect of line-tying can be determined.

(iii) Gravitational effects. Strangely enough there are very few detailed calculations that actually include gravity. The work of Zweibel (1981), Schindler *et al* (1983), Hood (1984), Melville *et al* (1986, 1987) and Migliuolo (1982) have all assumed isothermal equlibria. More detailed calculations are required for typical prominence fields and temperature profiles.

REFERENCES

Amari, T and Aly, J J: 1988, *Astron. Astrophys.*, in press.

An, C-H: 1982, *Solar Phys.* **75**, 19.

An, C-H: 1984, *Astrophys. J.* **276**, 755.

An, C-H: 1984, *Astrophys. J.* **281**, 419.

An, C-H: 1985, *Astrophys. J.* **298**, 409.

An, C-H: 1986, *Astrophys. J.* **304**, 532.

An, C-H: 1987, *Coronal and Prominence Plasmas*, NASA Conference Publication 2442 (ed. A Poland), p.41.

An, C-H, Bao, J J and Wu, S T: 1986, *Coronal and Prominence Plasmas*, NASA Conference Publication 2442 (ed. A Poland), p.51.

An, C-H, Bao, J J, Wu, S T and Suess, S T: 1987, submitted to Solar Phys.

An, C-H, Suess, S T, Tandberg-Hanssen, E and Steinolfson, R S: 1985, *Solar Phys.* **102**, 165.

An, C-H, Wu, S T and Bao, J J: 1988, in Proc. Mallorca Workshop on "Dynamics and Structure of Solar Prominences" (eds J L Ballester and E R Priest).

Antiochos, S: 1979, *Astrophys. J.* **232**, L125.

Anzer, U: 1968, *Solar Phys.* **3**, 298.

Anzer, U: 1969, *Solar Phys.* **8**, 37.

Anzer, U: 1972, *Solar Phys.* **24**, 324.

Anzer, U: 1979, IAU Coll. No. 44, p.322 (ed. E Jensen, P Maltby, F Q Orrall).

Anzer, U: 1984, *Measurements of Solar Vector Magnetic Fields*, (ed M J Hagyard) NASA Conf. Pub. 2374, p.101.

Anzer, U and Pneuman, G W: 1982, *Solar Phys.* **79**, 129.

Anzer, U and Priest, E R: 1985, *Solar Phys.* **95**, 263.

Apushkinskij, G P and Topchilo, N A: 1984, *Astron. Z.* **61**, 1150.

Arnaud, J: 1982a, *Astron. Astrophys.* **112**, 350.

Arnaud, J: 1982b, *Astron. Astrophys.* **116**, 248.

Arnaud, J and Newkirk, G: 1987, *Astron. Astrophys.* **178**, 263.

Athay, R G: 1966, *Astrophys. J.* **145**, 784.

Athay, R G: 1972, *The Solar Chromosphere and Corona*, D Reidel Publ. Comp. Dordrecht, Holland.

Athay, R G: 1985, *Solar Phys.* **100**, 257.

Athay, R G, Gurman, J B, Henze, N and Shine, R A: 1983, *Astrophys. J.* **265**, 219.

Athay, R G and Illing, R M E: 1986, *J. Geophys. Res.* **91**, 10961.

Athay, R G, Jones, H and Zirin, H: 1985, *Astrophys. J.* **288**, 363.

Athay, R G, Klimchuck, J A, Jones, H P and Zirin, H: 1986, *Astrophys. J.* **303**, 884.

Athay, R G, Low, B C and Rompolt, B: 1987, *Solar Phys.* **110**, 359.

Athay, R G, Querfeld, C W, Smartt, R N, Landi degl'Innocenti, E and Bommier, V: 1983, *Solar Phys.* **89**, 3.

Babcock, H W and Babcock, H D: 1955, *Astrophys. J.* **121**, 349.

Ballester, J L: 1984, *Solar Phys.* **94**, 151.

Ballester, J L and Kleczek, J: 1983, *Solar Phys.* **87**, 261.

Ballester, J L and Kleczek, J: 1984, *Solar Phys.* **90**, 37.

Ballester, J L and Priest, E R: 1987, *Solar Phys.* **109**, 335.

Ballester, J L and Priest, E R: 1988, Proc. Workshop on Dynamics and Structure of Solar Prominences, Universitat de les Illes Balears, in press.

Balthasar, H, Knolker, M, Stellmacher, G and Wiehr, E: 1986, *Astron. Astrophys.* **163**, 343.

Bashkirtsev, V S: 1975, *Issl. po Geom. Aeron. i Fis. Soln.* **37**, 7.

Bashkirtsev, V S, Kobanov, N I and Mashnich, G P: 1987, *Solar Phys.* **109**, 399.

Bashkirtsev, V S and Mashnich, G P: 1976, *Soln. Dann.* **7**, 82.

Bashkirtsev, V S and Mashnich, G P: 1980, *Phys. Solariterr. Potdsdam* **13**, 118.

Bashkirtsev, V S and Mashnich, G P: 1984, *Solar Phys.* **91**, 93.

Bashkirtsev, V S and Mashnich, G P: 1987, SibIZMIR preprint, **7-87**.

REFERENCES

Bateman, G: 1980, *MHD Instabilities*, MIT Press.

Baur, T G, Elmore, D E, Lee, R H, Querfeld, C W and Rogers, S R: 1981, *Solar Phys.* **70**, 395.

Baur, T G, House, L L and Hull, H K: 1980, *Solar Phys.* **65**, 111.

Beckers, J M: 1964, Thesis, Utrecht University.

Bernstein, I B, Frieman, E A, Kruskal, M D and Kulsrud, R M: 1958, *Proc. Roy. Soc. London* **A244**, 17.

Bessey, R J and Liebenberg, D H: 1984, *Solar Phys.* **94**, 239.

Billings, D E and Kober, C: 1957, *Sky Telescope* **17**, 63.

Birn, J, Goldstein, H and Schindler, K: 1978, *Solar Phys.* **57**, 81.

Bohlin, J D: 1970, *Solar Phys.* **13**, 153.

Bommier, V: 1977, These de Doctorat de 3eme Cycle, Universite de Paris 6.

Bommier, V: 1986, *Coronal and Prominence Plasmas*, NASA Conference Publication 2442 (ed. A Poland), p. 209.

Bommier, V: 1987, These de Doctorat d"Etat, Universite de Paris 7.

Bommier, V, Leroy, J L and Sahal-Brechot, S: 1981, *Astron. Astrophys.* **100**, 231.

Bommier, V, Leroy, J L and Sahal-Brechot, S: 1985, NASA Conference Publication 2374, p.375, (ed. M J Hagyard).

Bommier, V, Leroy, J L and Sahal-Brechot, S: 1986a, *Astron. Astrophys.* **156**, 79.

Bommier, V, Leroy, J L and Sahal-Brechot, S: 1986b, *Astron. Astrophys.* **156**, 90.

Bommier, V and Sahal-Brechot, S: 1979, IAU Coll. No. 44, p.87, (eds. E Jensen, P Maltby, F Q Orrall)

Bonnet, R M, Brunner, E C, Acton, L W, Brown, W A and Decaudin, M: 1980, *Astrophys. J.* **237**, L47.

Browning, P K and Priest, E R: 1986, *Solar Phys.* **106**, 335.

Bruzek, A and Kuperus, M: 1972, *Solar Phys.* **24**, 3.

Bumba, V and Howard, R: 1965, *Astrophys. J.* **141**, 1502.

Cargill, P, Hood, A W and Migliuolo, S: 1986, *Astrophys. J.* **309**, 402.

Chapman, R D: 1981, *Solar Phys.* **71**, 151.

Charvin, P: 1965, *Ann. Astrophys.* **28**, 877.

Chiuderi, C and Van Hoven, G: 1979, *Astrophys. J.* **232**, L69.

Chiuderi, C, Einaudi, G, Toricelli-Campioni, G: 1981, *Astron. Astrophys.* **97**, 27.

Connor, J W, Hastie, R J and Taylor, J B: 1979, *Proc. Roy. Soc. London* **A365**, 1.

Cowling, T G: 1976, *Magnetohydrodynamics*, 2nd edn, Adam Hilger, Bristol, UK.

Cox, D P and Tucker, W H: 1969, *Astrophys. J.* **157**, 1157.

Cram, L E: 1986, *Astrophys. J.* **300**, 830.

Cui Lian-Shu, Hu Ju, Ji Guo-Ping, Ni Xiang-Bin, Huang You-Ran and Fang Cheng: 1985, *Chin. Astron. Astrophys.* **9**, 49.

Davis, J M and Krieger, S: 1982, *Solar Phys.* **81**, 325.

D'Azambuja, L and D'Azambuja, M: 1948, *Ann. Obs. Paris. Meudon* **6**, 7.

De Bruyne, P and Hood, A W: 1988, *Solar Phys.*, in press.

Demoulin, P and Einaudi, G: 1988, in Proc. Mallorca Workshop on "Dynamics and Structure of Solar Prominences", (eds J L Ballester and E R Priest).

Demoulin, P and Priest, E R: 1988, *Astron. Astrophys.*, in press.

Demoulin, P, Raadu, M A, Malherbe, J M and Schmieder, B: 1987, *Astron. Astrophys.* **183**, 142.

Dere, K P, Bartoe, J-D F and Brueckner, G E: 1986, *Astrophys. J.* **310**, 456.

Dewar, R and Glasser, A: 1983, *Phys. Fluids* **26**, 3039.

Dungey, J W: 1953, *Mon. Not. Roy. Astron. Soc.* **113**, 180.

Dunn, R B: 1960, Ph.D. Thesis, Harvard University.

Dupree, A K: 1972, *Astrophys. J.* **178**, 527.

Duvall, T L, Wilcox, J M, Svalgaard, L, Scherrer, P H and McIntosh, P S: 1977, *Solar Phys.* **55**, 63.

Eddy, J A: 1973, *Solar Phys.* **30**, 385.

Einaudi, G, Toricelli-Campioni, G and Chiuderi, C: 1984, *Solar Phys.* **92**, 99.

REFERENCES

Einaudi, G and Van Hoven, G: 1981, *Phys. Fluids* **24**, 1092.

Einaudi, G and Van Hoven, G: 1983, *Solar Phys.* **88**, 163.

Engvold, O: 1976, *Solar Phys.* **49**, 283.

Engvold, O: 1981, *Solar Phys.* **70**, 315.

Engvold, O: 1987, *Solar and Stellar Coronal Structure and Dynamics* 17-20 August, Sacramento Peak, New Mexico (in press).

Engvold, O and Brynildsen, N: 1986, *Coronal and Prominence Plasmas*, NASA Conference Publication 2442 (ed. A Poland), p. 97.

Engvold, O and Jensen, E: 1977, *Solar Phys.* **52**, 37.

Engvold, O, Jensen, E and Andersen, B W: 1979, *Solar Phys.* **62**, 331.

Engvold, O and Keil, S: 1986 *Coronal and Prominence Plasmas*, NASA Conference Publication 2442 (ed. A Poland), p. 169.

Engvold, O, Kjeldseth-Moe, O, Bartoe, J-D F and Brueckner, G E: 1987, 21st ESLAB Symposium, June 22-25, Bolkesjoe, Norway. ESA SP-275, 21.

Engvold, O and Malville, J M: 1977, *Solar Phys.* **52**, 369.

Engvold, O, Malville, J M and Livingstone, W: 1978, *Solar Phys.* **60**, 57.

Engvold, O, Tandberg-Hanssen, E and Reichmann, E: 1985, *Solar Phys.* **96**, 35.

Fang, C, Zhang, Q Z, Yin, S Y and Livingstone, W: 1988, submitted to *Scientia Sinica*.

Field, G B: 1965, *Astrophys. J.* **142**, 531.

Fontenla, J M and Rovira, M: 1985, *Solar Phys.* **96**, 53.

Forbes, T G and Malherbe, J M: 1986, *Astrophys. J. Letts* **302**, L67.

Forbes, T G, Malherbe, J M and Priest, E R: 1988, *Solar Phys.*, in press.

Foukal, P V: 1971, *Solar Phys.* **19**, 59.

Foukal, P V: 1976, *Astrophys. J.* J**210**, 575.

Frazier, E N: 1972, *Solar Phys.* **24**, 98.

Furth, H P, Killean, J and Rosenbluth, M: 1963, *Phys. Fluids*, **6**, 459.

Gaizauskas, V: 1985, Proc. Workshop on Solar Physics and Interplanetary Travelling Phenomena, (eds C de Jager and Chen Biao), 710.

Galindo-Trejo, J and Schindler, K: 1984, *Astrophys. J.* **277**, 422.

Galloway, D and Jones, C: 1988, *Astrophys. J.*, in press.

Gary, D: 1986, *Coronal and Prominence Plasmas*, NASA Conference Publication 2442 (ed. A Poland), p. 121.

Gilman, P A and Miller, J: 1986, *Astrophys. J. Suppl.* **256**, 316.

Giovanelli, R G: 1982, *Solar Phys.* **77**, 27.

Glatzmaier, G A: 1985, *Astrophys. J.* **291**, 300.

Gnevyshev, M N and Makarov, V I: 1985, *Solar Phys.* **95**, 189.

Goedbloed, J: 1983, *Lecture Notes on Ideal Magnetohydrodynamics*, Rijnhuizen Report 83-145, Netherlands.

Grossman-Doerth and Von Uexkull, M: 1971, *Solar Phys.* **20**, 31.

Gu, X O, Ding, J P and Tang, Y H: 1984, *Scientia Sinica* **973**.

Habbal, S and Rosner, R: 1980, *Astrophys. J.* **234**, 1113.

Hanaoka, Y and Kurokawa, H: 1986, *Solar Phys.* **105**, 133.

Hanle, W: 1924, *Z. Phys.* **30**, 93.

Hansen, S F and Hansen, R T: 1975, *Solar Phys.* **44**, 503.

Harvey, J W: 1969, Ph.D. thesis, University of Colorado.

Heasley, J N and Mihalas, D: 1976, *Astrophys. J.* **205**, 273.

Heasley, J N, Mihalas, D and Poland, A: 1974, *Astrophys. J.* **192**, 181.

Heasley, J N and Milkey, R W: 1976, *Astrophys. J.* **210**, 827.

Heasley, J N and Milkey, R W: 1978, *Astrophys. J.* **221**, 677.

Heinzel, P: 1987, 10th European Region Meeting, IAU, Prague, Tchechoslovaky.

Heinzel, P, Gouttebroze, P and Vial, J C: 1987, *Astron. Astrophys.* **183**, 351.

Heinzel, P, Gouttebroze, P and Vial, J C: 1988, in Proc.Mallorca Workshop on "Dynamics and Structure of Solar Prominences" (eds J L Ballester and E R Priest).

Heinzel, P and Karlicky, M: 1987, *Solar Phys.* **110**, 343.

Heinzel, P and Rompolt, B: 1987, *Solar Phys.* **110**, 171.

Hermans, L and Martin, S F: 1986, *Coronal and Prominence Plasmas*, NASA Conference Publication 2442 (ed. A Poland), p. 369.

Heyvaerts, J: 1974, *Astron. Astrophys.* **37**, 65.

Hiei, E, Ishiguro, M, Kosugi, T and Shibasaki, K: 1986, *Coronal and Prominence Plasmas*, NASA Conference Publication 2442 (ed. A Poland), p. 109.

Hiei, E and Widing, K G: 1979, *Solar Phys.* **61**, 407.

Hildner, E: 1974, *Solar Phys.* **35**, 123.

Hirayama, T: 1971, *Solar Phys.* **17**, 50.

Hirayama, T: 1979, IAU Coll. No.44, p.4 (ed. E Jensen, P Maltby, F Q Orrall).

Hirayama, T: 1985, *Solar Phys.* **100**, 415.

Hirayama, T: 1986, *Coronal and Prominence Plasmas*, NASA Conference Publication 2442 (ed. A Poland), p. 149.

Hood, A W: 1983, *Solar Phys.* **87**, 279.

Hood, A W: 1984, *Geophys. Astrophys. Fluid Dynamics* **28**, 223.

Hood, A W: 1986a, *Solar Phys.* **103**, 329.

Hood, A W: 1986b, *Solar Phys.* **105**, 307.

Hood, A W and Anzer, U: 1987, *Solar Phys.* **111**, 333.

Hood, A W and Anzer, U: 1988, *Solar Phys.*, in press.

Hood, A W and Priest, E R: 1979, *Solar Phys.* **64**, 303.

Hood, A W and Priest, E R: 1979, *Astron. Astrophys.* **77**, 233.

Hood, A W and Priest, E R: 1980a, *Solar Phys.* **66**, 113.

Hood, A W and Priest, E R: 1980b, *Astrophys. J.* **87**, 126.

Hood, A W and Priest, E R: 1981, *Geophys. Astrophys. Fluid Dynamics* **17**, 297.

House, L L: 1971, IAU Symp. No.43, p.130 (ed. R Howard), D Reidel, Dordrecht, Holland.

Howard, R and Harvey, J W: 1964, *Astrophys. J.* **139**, 1328.

Hyder, C L: 1965a, *Astrophys. J.* **141**, 1374.

Hyder, C L: 1965b, *Astrophys. J.* **141**, 272.

Hyder, C L: 1966, *Z. f. Ap.* **63**, 78.

Illing, R M E and Athay, G: 1986, *Solar Phys.* **105**, 173.

Illing, R M E and Hundhausen, A J: 1986, *J. Geophys. Res.* **91**, 10951.

Ioshpa, B A, Kozhevatov, I E, Kulikova, E K and Mogilevskij, E I: 1986, *Soln. Dannye* **5**, 68.

Ishimoto, K and Kurokawa, H: 1984, *Solar Phys.* **93**, 105.

Jensen, E: 1986, *Coronal and Prominence Plasmas*, NASA Conference Publication 2442 (ed. A Poland), p. 63.

Jensen, E, Maltby, P and Orrall, F Q (eds): 1979, *Physics of Solar Prominences*, IAU Colloq. **44**.

Jordan, C and Wilson, R: 1971, in *Physics of the Solar Corona*, (ed. C J Macris), Dordrecht: Reidel, p. 219.

Joselyn, J A, Munroe, R H and Holzer, T E: 1979, *Astrophys. J. Suppl. Ser.* **40**, 793.

Kaastra, J S: 1985, Ph.D. Thesis, Rijkuniversiteit-Utretcht-Nederlands.

Kahler, S, Cliver, E W, Cane, H V, McGuire, R E, Stone, R G and Sheeley, N R: 1986, *Astrophys. J.* **302**, 504.

Kahler, S and Kreplin, R: 1970, *Solar Phys.* **14**, 372.

Kawaguchi, I: 1967, *Solar Phys.* **1**, 420.

Kiepenheuer, K O: 1953, *The Sun* p.322 (ed. G P Kuiper), The University of Chicago Press.

Kim, I S, Klepikov, V I, Koutchmy, S, Stepanov, A I and Stellmacher, G: 1987, IZMIRAN preprint No.50 (739).

Kim, I S, Koutchmy, S, Stellmacher, G and Stepanov, A I: 1988, *Role of Fine-Scale Magnetic Fields on the Structure of the Solar Atmosphere* (eds E Schroter, M Vazquez, A Wyller) Camb. Univ. Press, p.289.

Kippenhahn, R and Schluter, R: 1957, *Zs. Ap.* **43**, 36.

Kjeldseth-Moe, O, Andreassen, O, Malt , P, Bartoe, J-D F, Brueckner, G E and Nicolas, K R: 1984, *Adv. Space Res.* **4**, 63.

REFERENCES

Kjeldseth-Moe, O, Cook, J W and Mango, S A: 1979, *Solar Phys.* **61**, 319.

Kleczek, J and Kuperus, M: 1969, *Solar Phys.* **6**, 72.

Klepikov, V and Platov, V: 1985, *Astron. Z.* **62**, 983.

Klimchuk, J A: 1986, *Coronal and Prominence Plasmas*, NASA Conference Publication 2442 (ed. A Poland), p. 183.

Kopp, R and Pneuman, G: 1976, *Solar Phys.* **50**, 85.

Koutchmy, S: 1987, Private communication.

Koutchmy, S, Lebecq, C and Stellmacher, G: 1983, *Astron. Astrophys.* **119**, 261.

Kubota, J: 1980, *Proc. Japan-France Seminar on Solar Phys.* (ed F Moriyama and J C Henoux) p.178.

Kubota, J, Tohmura, I and Uesugi, A: 1988, in Proceedings *Vistas in Astronomy* of the 3rd Asian Pacific, Regional Meeting of IAU (Beyin, China), Oct 1987.

Kubota, J and Uesugi, A: 1986, *Publ. Astron. Soc. Japan* **38**, 903.

Kuin, N P and Martens, P: 1986, *Coronal and Prominence Plasmas*, NASA Conference Publication 2442 (ed. A Poland), p. 241.

Kundu, M R: 1986, *Coronal and Prominence Plasmas*, NASA Conference Publication 2442 (ed. A Poland), p. 117.

Kundu, M R, Gaizauskas, V, Woodgate, V E, Schmahl, E J, Shine, R and Jones, H P: 1985, *Astrophys. J. Suppl.* **621**.

Kuperus, M and Raadu, M: 1974, *Astron. Astrophys.* **31**, 185.

Kuperus, M and Raadu, M: 1974, *Astrophys. J.* **31**, 189.

Kuperus, M and Tandberg-Hansen, E: 1967, *Solar Phys.* **2**, 39.

Kuperus, M and Van Tend, W: 1981, *Solar Phys.* **71**, 125.

Kurokawa, H, Hanaoka, Y, Shibata, K and Uchida, Y: 1987, *Solar Phys.* **108**, 251.

Lamb, F K: 1970, *Solar Phys.* **12**, 186.

Landi degl'Innocenti, E: 1982, *Solar Phys.* **79**, 291.

Landi degl'Innocenti, E, Bommier, V and Sahal-Brechot, S: 1987, *Astron. Astrophys.* **186**, 335.

Landman, D A: 1984, *Astrophys. J.* **279**, 438.

Landman, D A: 1985a, *Astrophys. J.* **295**, 220.

Landman, D A: 1985b, *Astrophys. J.* **290**, 369.

Laval, G, Mercier, C and Pellat, R: 1965, *Nuclear Fusion* **5**, 156.

Lee, R H, Rust, D M and Zirin, H: 1965, *Applied Optics* **4**, 1081.

Lerche, I and Low, B C: 1977, *Solar Phys.* **53**, 385.

Leroy, J L: 1979, in Jensen et al (1979), p.56.

Leroy, J L: 1985, NASA Conference Publication 2374, p.121 (ed. M J Hagyard).

Leroy, J L: 1987, *Solar and Stellar Coronal Structure and Dynamics*, 17-20 August, Sacramento Peak, New Mexico, in press.

Leroy, J L: 1988, Proc. 9th Sacramento Peak Summer Meeting, (ed. R C Altrock), in press.

Leroy, J L: 1988, Proc. Workshop on "Dynamics and Structure of Solar Prominences", (ed. J L Ballester and E R Priest), in press.

Leroy, J L, Bommier, V and Sahal-Brechot, S: 1983, *Solar Phys.* **83**, 135.

Leroy, J L, Bommier, V and Sahal-Brechot, S: 1984, *Astron. Astrophys.* **131**, 33.

Leroy, J L and Servajan, R: 1966, *Ann. Astrophys.* **29**, 263.

Liggett, M and Zirin, H: 1984, *Solar Phys.* **91**, 259.

Lites, B W, Bruner, E C, Chipman, E G, Shine, R A, Rottman, G J, White, O R and Athay, R G: 1976, *Astrophys. J.* **210**, L111.

Loughead, R E, Jia-Long Wang and Blows, G: 1983, *Astrophys. J.* **274**, 883.

Low, B C: 1975, *Astrophys. J.* **198**, 211.

Low, B C: 1982, *Rev. Geophys. Space Phys.*, **20**, 145.

Low, B C: 1985, *Solar Phys.* **100**, 809.

Low, B C, Hundhausen, A J and Zweibel, E G: 1983, *Phys. Fluids* **26**, 2731.

Lyot, B: 1934, *C. R. Acad. Sci.* **198**, 249.

McClymont, A and Craig, I: 1985, *Astrophys. J.* **289**, 820.

McIntosh, P S: 1972a, *Solar Activity Observation and Predictions*, p.65; ed. P S McIntosh and M Dryer, The MIT Press, Cambridge, Mass.

REFERENCES

McIntosh, P S: 1972b, *Rev. Geophys. and Space Phys.* **10**, 837.

McIntosh, P S: 1980, IAU Symp. No.91, p.25 (ed. M Dryer and E Tandberg-Hanssen), D. Reidel Publ. Comp., Dordrecht, Holland.

MacQueen, R M: 1988, private communication.

MacQueen, R M, Eddy, J A, Gosling, J T, Hildner, E, Munroe, R H, Newkirk, G A, Poland, A I and Ross, C L: 1974, *Astrophys. J.* **187**, L85.

MacQueen, R M, Sime, D G and Picat, J-P: 1983, *Solar Phys.* **83**, 103.

Makarov, V I: 1984, *Solar Phys.* **93**, 393.

Maksimov, V P and Ermakova, L V: 1985, *Astr. Zhurnal* **62**, 558.

Malherbe, J M: 1987, these de Doctorat d'Etat, Universite de Paris VII.

Malherbe, J M: 1987, IAU report on the Physics of Solar Prominences and Filaments.

Malherbe, J M and Forbes, T: 1986, *Coronal and Prominence Plasmas*, NASA Conference Publication 2442 (ed. A Poland), p. 225.

Malherbe, J M and Priest, E R: 1983, *Astron. Astrophys.* **123**, 80.

Malherbe, J M, Schmieder, B and Mein, P: 1981, *Astron. Astrophys.* **55**, 103.

Malherbe, J M, Schmieder, B, Ribes, E and Mein, P: 1983, *Astron. Astrophys.* **119**, 197.

Malherbe, J M, Schmieder, B, Mein, P and Tandberg-Hanssen, E: 1987, *Astron. Astrophys.* **172**, 316.

Malville, J M: 1979a, IAU Coll. **44**, 192 (ed. E Jensen, P Maltby, F Q Orrall).

Malville, J M: 1979b, IAU Coll. **44**, 214 (ed. E Jensen, P Maltby, F Q Orrall).

Malville, J M and Schindler, M: 1981, *Solar Phys.* **70**, 115.

Mariska, J T: 1986, *Ann. Rev. Astron. Astrophys.* **24**, 23.

Mariska, J T, Doscheck, G A and Feldman, U: 1979, *Astrophys. J.* **232**, 929.

Martin, S F: 1973, *Solar Phys.* **31**, 3.

Martin, S F: 1980, *Solar Phys.* **68**, 217.

Martin, S F: 1986, *Coronal and Prominence Plasmas*, NASA Conference Publication 2442 (ed. A Poland), p. 73.

Martin, S F, Deszo, L, Gesztelyi, L, Antalova, A, Kucra, A and Harvey, K L: 1983, *Adv. Space Res.* **2**, 11, 39.

Martres, M J: 1988, private communication.

Martres, M J, Mein, P, Schmieder, B and Soru-Escaut, I: 1981, *Solar Phys.* **69**, 301.

Martres, M J, Mouradian, Z, Ribes, E and Soru-Escaut, I: 1987 Proc. of Mallorca Workshop on "Dynamics and Structure of Solar Prominences" , (Eds J L Ballester and E R Priest).

Martres, M J, Rayrole, J and Soru-Escaut, I: 1976, *Solar Phys.* **46**, 137.

Martres, M J, Rayrole, J, Semel, M, Soru-Escaut, I, Tanaka, K, Makita, M, Moriyama, F and Unno, W: 1982, *Publ. Astr. Soc. Japan* **34**, 299.

Mein, P: 1977 *Solar Phys.* **54**, 45.

Mein, P and Mein, N: 1982 *Solar Phys.* **80**, 161.

Mein, P and Mein, N: 1988, *Astron. Astrophys.*, in press.

Mein, P, Mein, N and Malherbe, J M: 1985, Theoretical problems in high resolution solar physics, Ed H V Schmidt, Proceedings of the MPA/LPARL Workshop in Munchen, 303.

Mein, P and Schmieder, B: 1988 Proc. Mallorca Workshop on "Dynamics and Structure of Solar Prominences", (eds J L Ballester and E R Priest).

Melville, J, Hood, A W and Priest, E R: 1986, *Solar Phys.* **105**, 291.

Melville, J, Hood, A W and Priest, E R: 1987, *Geophys. Astrophys. Fluid Dynamics* **39**, 83.

Menzel, D H: 1951, Proceedings of Conferences on Dynamics of Ionized Media, London.

Migliuolo, S: 1982, *J. Geophys. Research* **87**, 8057.

Migliuolo, S, Cargill, P and Hood, A W: 1984, *Astrophys. J.* **281**, 413.

Mihalas, D: 1978, *Stellar Atmospheres*, W H Freeman and Company, Second Edition.

Milne, A M, Priest, E R and Roberts, B: 1979, *Astrophys. J.* **232**, 304.

Mok, Y and Van Hoven, G: 1982, *Phys. Fluids* **25**, 636.

Morozhenko, N N: 1978, *Solar Phys.* **58**, 47.

Mouradian, Z and Leroy, J I: 1977, *Solar Phys.* **51**, 103.

Mouradian, Z and Martres, M J: 1986, *Coronal and Prominence Plasmas*, NASA Conference Publication 2442 (ed. A Poland), p. 221.

REFERENCES

Mouradian, Z, Martres, M, Soru-Escaut, I: 1980, *Proc. Japan-France Seminar on Solar Phys.* (ed F Moriyama and J C Henoux), p. 195.

Mouradian, Z, Martres, M, Soru-Escaut, I and Gesztelyi, L: 1987, *Astron. Astrophys.* **183**, 129.

Mouradian, Z, Martres, M and Soru-Escaut, I: 1987, *Coronal and Prominence Plasmas*, NASA Conference Publication 2442 (ed. A Poland), p.221.

Nakagawa, Y and Malville, J M: 1969, *Solar Phys.* **9**, 102.

Newcomb, W: 1960, *Annals of Physics* **10**, 232.

Newkirk, G A: 1967, *Ann. Rev. Astron. Astrophys.* **5**, 213. *Solar Phys.* **49**, 57.

Nicolas, K R, Kjeldseth-Moe, O, Bartoe, J-D F and Brueckner, G E: 1982, *Solar Phys.* **81**, 253.

Nikolsky, G M, Kim, I S and Koutchmy, S: 1982, *Solar Phys.* **81**, 81.

Nikolsky, G M, Kim, I S, Koutchmy, S and Stellmacher, G: 1984, *Astron. Astrophys.* **140**, 112.

Nikolsky, G M, Kim, I S, Koutchmy, S, Stepanov, A I and Stellmacher, G: 1985, *Sov. Astronomy* **29**, 669.

Noens, J C: 1987, in Proceedings of Sac Peak Meeting, (ed. R C Altrock), in press.

Noyes, R, Dupree, A, Huber, M C, Parkinson, W, Reeves, E and Withbroe, G: 1972, *Astrophys. J.* **178**, 515.

Ohman, Y: 1929, *Mon. Not. Roy. Astron. Soc.* **89**, 479.

Orrall, F Q and Schmahl, E J: 1976, *Solar Phys.* **50**, 365.

Osherovich, V: 1985, *Astrophys. J.* **297**, 314.

Parker, E: 1953, *Astrophys. J.* **117**, 431.

Parker, E N: 1963, *Astrophys. J. Suppl.* ser., **8**, 177.

Petschek, H E: 1964, NASA Publication **50**, p.425.

Pikel'ner, S B: 1971, *Solar Phys.* **17**, 44.

Ploceniak, S and Rompolt, B: 1973, *Solar Phys.* **29**, 399.

Pneuman, G W: 1968, *Solar Phys.* **3**, 578.

Pneuman, G W: 1973, *Solar Phys.* **28**, 249.

Pneuman, G W: 1980, *Solar Phys.* **65**, 369.

Pneuman, G W: 1983, *Solar Phys.* **88**, 219.

Pneuman, G W: 1984, *Solar Phys.* **94**, 299.

Pneuman, G W and Kopp, R: 1971, *Solar Phys.* **18**, 258.

Pneuman, G W and Kopp, R: 1978, *Solar Phys.* **57**, 49.

Poland, A I: 1978, *Solar Phys.* **57**, 141.

Poland, A I (ed): 1986, *Coronal and Prominence Plasmas*, NASA Conf. Pub. 2442.

Poland, A I and Anzer, U: 1971, *Solar Phys.* **19**, 401.

Poland, A I and MacQueen, R M: 1981, *Solar Phys.* **71**, 361.

Poland, A I and Mariska, J T: 1986, *Solar Phys.* **104**, 303.

Poland, A I, Mariska, J T and Klimchuk, J A: 1986, *Coronal and Prominence Plasmas*, NASA Conference Publication 2442 (ed. A Poland), p. 57.

Poland, A I, Schumanich, A, Athay, R G and Tandberg-Hanssen, E: 1971, *Solar Phys.* **391**.

Poland, A I and Tandberg-Hanssen, E: 1983, *Solar Phys.* **84**, 63.

Priest, E R: 1982, *Solar Magnetohydrodynamics*, D Reidel, Dordrecht.

Priest, E R: 1984, *Magnetic Reconnection in Space and Lab Plasmas*, (ed. E W Hones), p. 63.

Priest, E R: 1985, *Rep. Prog. Phys.* **48**, 955.

Priest, E R: 1986, *Solar Phys.* **104**, 1.

Priest, E R: 1987, *Role of Fine-Scale Magnetic Fields on the Structure of the Solar Atmosphere* (ed. E Schroter, M Vasquez, A Wyller) Camb. Univ. Press. p. 297.

Priest, E R, Anzer, U and Hood, A W: 1988, submitted.

Priest, E R and Forbes, T G: 1986, *J. Geophys. Res.* **91**, 5579.

Priest, E R and Smith, E A: 1979, *Solar Phys.* **64**, 217.

Prokakis, T and Tsiropoula, G: 1987, Proc. Mallorca Workshop on "Dynamics and Structure of Prominences", (eds. J L Ballester and E R Priest).

Querfeld, C W, Smartt, R N, Bommier, V, Landi degl'Innocenti, E and House, L L: 1985, *Solar Phys.* **96**, 277.

Raadu, M: 1972, *Solar Phys.* **22**, 425.

REFERENCES

Raadu, M A and Kuperus, M: 1973, *Solar Phys.* **28**, 77.

Raadu, M A, Malherbe, J M, Schmieder, B and Mein, P: 1987, *Solar Phys.* **109**, 59.

Raadu, M A, Schmieder, B, Mein, N and Gesztelyi, L: 1988, *Astron. Astrophys.*, in press.

Rabin, D: 1986, *Coronal and Prominence Plasmas*, NASA Conference Publication 2442 (ed. A Poland), p. 135.

Ratier, G: 1975, *Nouv. Revue Optique* **6**, 149.

Raymond, J C and Doyle, J G: 1981, *Astron. Astrophys. J.* **247**, 686.

Rayrole, J: 1981, in Proceedings of the Japan-France Seminar on Solar Physics, (ed. Moriyama and Henoux), 258.

Ribes, E: 1986, *C. R. Acad. Sc. Paris* **302**, serie II, 14.

Ribes, E and Laclare, F: 1988, *Geophys. and Astrophys. Fluid Dynamics* **41**, 171.

Ribes, E, Mein, P and Mangeney, A: 1985, *Nature* **318**, 170.

Ribes, E and Unno, W: 1980, *Astron. Astrophys.* **91**, 129.

Roberts, B: 1984, in *Hydromagnetics of the Sun*, ESA SP-220, p. 137.

Roberts, B: 1985, Ch 3 of *Solar System Magnetic Fields* (ed E R Priest), D Reidel, Dordrecht.

Roberts, P H: 1967, *An Introduction to Magnetohydrodynamics*, Longmans, London.

Rompolt, B: 1986, *Coronal and Prominence Plasmas*, NASA Conference Publication 2442 (ed. A Poland), p. 81.

Rosner, R, Low, B C and Holzer, T: 1984, in *Physics of the Sun*, Reidel.

Rust, D M: 1966, NCAR Cooperative Thesis No.6, Univ. of Colorado.

Rust, D M: 1967, *Astrophys. J.* **150**, 313.

Rust, D M: 1970, *Astrophys. J.* **160**, 315.

Rust, D M: 1972, Air Force Survey in Geophysics No.237.

Rust, D M: 1979, IAU Coll. No.44, p.172 (ed. E Jensen, P Maltby, F Q Orrall).

Rust, D M: 1984, *Solar Phys.* **93**, 73.

Sahal-Brechot, S: 1981, *Space Sci. Rev.* **30**, 99.

Sahal-Brechot, S: 1984, *Ann. Phys. Fr.* **9**, 705.

Sahal-Brechot, S, Bommier, V and Leroy, J L: 1977, *Astron. Astrophys.* **59**, 223.

Sahal-Brechot, S, Malinovsky, M and Bommier, V: 1986, *Astron. Astrophys.* **168**, 284.

Saito, K and Hyder, C L: 1968, *Solar Phys.* **5**, 61.

Saito, K and Tandberg-Hanssen, E: 1973, *Solar Phys.* **31**, 105.

Sastri, J H: 1987, *Solar Phys.* **105**, 191.

Schindler, K, Birn, J and Janicke, L: 1983, *Solar Phys.* **87**, 103.

Schmahl, E J: 1979, IAU Coll. 44 (eds: E Jensen, P Maltby and F Q Orrall), p.102.

Schmahl, E J, Foukal, P V, Huber, M C, Noyes, R W, Reves, E M, Timothy, J G, Vernazza J E and Withbroe, G L: 1974, *Solar Phys.* **39**, 337.

Schmahl, E J, Mouradian, Z, Martres, M J and Soru-Escaut, I: 1982, *Solar Phys.* **77**, 121.

Schmahl, E J and Orrall, F Q: 1980, *Astrophys. J.* **240**, 908.

Schmahl, E J and Orrall, F Q: 1986, *Coronal and Prominence Plasmas*, NASA Conference Publication 2442 (ed. A Poland), p. 127.

Schmieder, B: 1988, Chapter 2 of this book.

Schmieder, B, Forbes, J G, Malherbe, J M and Machado, M E: 1987, *Astrophys. J.* **317**, 956.

Schmieder, B, Malherbe, J M, Mein, P and Tandberg-Hanssen, E: 1984a, *Astron. Astrophys.* **136**, 81.

Schmieder, B, Malherbe, J M, Poland, A I and Simon, G: 1985b, *Astron. Astrophys.* **153**, 64.

Schmieder, B, Malherbe, J M, Mein, P and Tandberg-Hanssen, E: 1986, *Coronal and Prominence Plasmas*, NASA Conference Publication 2442 (ed. A Poland), p. 197.

Schmieder, B, Poland, A I, Thompson, B and Demoulin, P: 1988, *Astron. Astrophys.*, in press.

Schmieder, B, Raadu, M A and Malherbe, J M: 1985a, *Astron. Astrophys.* **142**, 249.

Schmieder, B, Ribes, E, Mein, P and Malherbe, J M: 1984b, *Mem. S.A. It.* **55**, 319.

Schroter, E H: 1986, *Solar Phys.* **100**, 141.

Secchi: 1875, *Le Soleil*, Library of the Observatoire de Paris.

Serio, S, Vaiana, G S, Godoli, G, Motta, S, Pirronello, V and Zappala, R A: 1978, *Solar Phys.* **59**,

REFERENCES

65.

She, Z S, Malherbe, J M and Raadu, M A: 1986, *Astron. Astrophys.* **164**, 364.

Simon, G, Mein, P, Vial, J C, Shine, R A and Woodgate, B E: 1982, *Astron. Astrophys.* **115**, 367.

Simon, G, Mein, N, Mein, P and Gesztelyi, L: 1984, *Solar Phys.* **93**, 325.

Simon, G, Schmieder, B, Demoulin, P and Poland, A I: 1986a, *Astron. Astrophys.* **166**, 319.

Simon, G, Gesztelyi, L, Schmieder, B and Mein, N: 1986b, *Coronal and Prominence Plasmas*, NASA Conference Publication 2442 (ed. A Poland), p. 229.

Smith, E A and Priest, E R: 1977 *Solar Phys.* **53**, 25.

Smith, S F and Ramsey, H R: 1967, *Solar Phys.* **2**, 158.

Smolkov, G Y: 1971, IAU Symp. No.43, p.710 (ed. R Howard), D Reidel, Dordrecht, Holland.

Snodgrass, H and Wilson, P: 1987, *Nature* **328**, 696.

Solovev, A A: 1985, *Soln. Dannye* **9**, 65.

Sonnerup, B U: 1970, *J. Plasma Phys.* **4**, 161.

Soru-Escaut, I, Martres, M J and Mouradian, Z: 1984, in Proc. "Solar and Terrestrial Predictions" Meeting, (ed. P Simon), T299, serie II, 9, 545.

Soru-Escaut, I, Martres, M J and Mouradian, Z: 1985, *Astron. Astrophys.* **145**, 19.

Soward, A M: 1982, *J. Plasma Phys.* **28**, 415.

Soward, A M and Priest, E R: 1982, *J. Plasma Phys.* **28**, 335.

Sparks, L and Van Hoven, G: 1985, *Solar Phys.* **97**, 283.

Sparks, L and Van Hoven, G: 1987, *Phys. Fluids* **26**, 2590.

Steele, C D C and Priest, E R: 1988, *Solar Phys.*, submitted.

Steinolfson, R S: 1983, *Phys. Fluids* **26**, 2590.

Steinolfson, R S and Van Hoven, G: 1984, *Astrophys. J.* **J276**, 391.

Stellmacher, G, Koutchmy, S and Lebecq, C: 1986, *Astron. Astrophys.* **162**, 307.

Stenflo, J O: 1973, *Solar Phys.* **32**, 41.

Stenflo, J O: 1985a, *Solar Phys.* **100**, 189.

Stenflo, J O: 1985b, NASA Conference Publication 2374, p.263 (ed. M J Hagyard).

Stepanov, V E: 1958, *Izv. Krymsk, Astrof. Obs.* **20**, 52.

Sturrock, P A and Smith, S M: 1968, *Solar Phys.* **5**, 87.

Suydam, B: 1958, IAEA Geneva Conf. **31**, 157.

Sweet, P A: 1958, in *Electromagnetic Phenomena in Cosmical Physics* (ed. B Lehnert), p. 123.

Tachi, T, Steinolfson, R S and Van Hoven, G: 1985, *Solar Phys.* **95**, 119.

Tandberg-Hanssen, E: 1970, *Solar Phys.* **15**, 359.

Tandberg-Hanssen, E: 1974, *Solar Prominences*, D Reidel, Dordrecht.

Tandberg-Hanssen, E: 1986, *Coronal and Prominence Plasmas*, NASA Conference Publication 2442 (ed. A Poland), p. 5.

Tandberg-Hanssen, E and Anzer, U: 1970, *Solar Phys.* **15**, 158.

Tandberg-Hanssen, E and Malville, J McKim: 1974, *Solar Phys.* **39**, 107.

Tang, F: 1986, *Solar Phys.* **105**, 2.

Tang, F: 1987, *Solar Phys.* **107**, 233.

Thiessen, G: 1951, *Z. f. Ap.* **30**, 8.

Tsubaki, T: 1988, Proc. 9th Sacramento Peak Summer Meeting, (ed R C Altrock) in press.

Tsubaki, T, Ohnishi, Y and Suematsu, Y: 1987, *Publ. Astron. Soc. Japan* **39**, 179.

Tsubaki, T and Takeuchi, A: 1986, *Solar Phys.* **104**, 313.

Uchida, Y: 1979, *Proceedings of the Japan-France seminar on Solar Phys.* (ed. J C Henoux), p. 169.

Van Hoven, G, Ma, S S and Einaudi, G: 1980, *Astron. Astrophys.* **97**, 232.

Van Hoven, G and Mok, Y: 1984, *Astrophys. J.* **282**, 267.

Van Hoven, G, Sparks, L and Tachi, T: 1986, *Astrophys. J.* **300**, 249.

Van Hoven, G, Tachi, T and Steinolfson, R S: 1984, *Astrophys. J.* **280**, 391.

Van Tend, W and Kuperus, M: 1978, *Solar Phys.* **59**, 115.

Velli, M and Hood, A W: 1986, *Solar Phys.* **106**, 353.

Velli, M and Hood, A W: 1987, *Solar Phys.* **109**, 351.

Velli, M and Hood, A W: 1988, *Solar Phys.*, in press.

Vial, J C: 1982a, *Astrophys. J.* **253**, 330.

Vial, J C: 1982b, *Astrophys. J.* **254**, 780.

Vial, J C: 1986, *Coronal and Prominence Plasmas*, NASA Conference Publication 2442 (ed. A Poland), p. 89.

Vial, J C, Gouttebroze, P, Artzner, G and Lemaire, P: 1979, *Solar Phys.* **61**, 39.

Vial, J C, Lemaire, P, Artzner, G and Gouttebroze, P: 1980, *Solar Phys.* **68**, 187.

Vizoso, G and Ballester, J L: 1987, Proc. Mallorca Workshop on "Dynamics and Structure of Prominences", (eds. J L Ballester and E R Priest).

Von Kluber, H: 1967, *Mon. Not. Roy. Astron. Soc.* **137**, 297.

Vrsnak, B: 1984, *Solar Phys.* **94**, 289.

Wagner, W: 1986, *Coronal and Prominence Plasmas*, NASA Conference Publication 2442 (ed. A Poland), p. 215.

Waldmeier, M: 1941, *Ergebnisse und Probleme der Sonnenforschung*, Leibzig, 234.

Waldmeier, M: 1957, *Z. f. Ap.* **42**, 34.

Waldmeier, M: 1972, *Solar Phys.* **27**, 143.

Wang, J L: 1985, *Scientia Sinica* **28**, 1308.

Wesson, J: 1981, *Plasma Physics and Nuclear Fusion Research*, Academic Press.

Wiehr, E: 1972, *Astron. Astrophys.* **18**, 79.

Wiehr, E, Stellmacher, G and Balthasar, H: 1984, *Solar Phys.* **94**, 285.

Wu, F: 1987, *Astrophys. J.*, **320**, 418.

Wu, F and Low, B C: 1987, *Astrophys. J.* **312**, 431.

Wu, S T, Bao, J J, An, C H and Tandberg-Hanssen, E: 1988, in Proc. Mallorca Workshop on "Dynamics and Structure of Solar Prominences", (eds. J L Ballester and E R Priest).

Wu, S T and Xiao, Y C: 1986, *Coronal and Prominence Plasmas*, NASA Conference Publication 2442 (ed. A Poland), p. 47.

Yang, C Y, Nicholls, R W and Morgan, F J: 1975, *Solar Phys.* **45**, 351.

Zarkhova, V V: 1984, *Astron. Astrofiz.* **51**, 45.

Zhang, Q Z and Fang, C: 1987, *Astron. Astrophys.* **175**, 277.

Zhang, Q Z, Livingston, W C, Hu, J and Fang, C: 1987, submitted to Solar Phys.

Zirin, H: 1961, *Astr. Zhurnal* **38**, 861.

Zirin, H: 1975, *Astrophys. J.* **199**, L63.

Zirin, H: 1988, private communication.

Zwaan, C: 1987, *Ann. Rev. Astron. Astrophys.* **25**, 83.

Zweibel, E: 1981, *Astrophys. J.* **249**, 731.

Zweibel, E: 1982, *Astrophys. J.* **258**, L53.

Zwingmann, W: 1987, *Solar Phys.* **111**, 309.

INDEX

Activation 7
Active prominence 1,62
Active-region (or plage) filament 1,3-4,7-8,96,136-137
Alfven speed 131
Alfven time 10,131-132,167
Alfven wave 11,37,154,158
    surface 172
Apparition brusque 117,124
Arcade, coronal 54,134,176-184
Arch, magnetic 50,52

Babcock magnetograph 77
Ballooning mode 177,180-181,188
Beta, plasma 10,74,138-139
Buoyancy, magnetic 43

Cancelling magnetic flux 42,55, 92,117
Catastrophe 176,182,189
Cavity, coronal 2,4,47-48,52-53, 57,68-73,93
Characteristics of prominences 15-22
Chromospheric injection 135-137
Circulation, meridional 23
Classes of prominences 1,143
Coalescence instability 132
Complex variable 146-147,151
Condensation 138-140
Conduction, thermal 63,130
Conduction time-scale 19,130
Continuity equation 9
Continuous spectrum 171
Corona, K- 52
Coronagraph 3-4,70-71,77,88
Coronal arcade 54,134,176-184
Coronal cavity 2,4,47-48,52-53, 57,68-73,93
Coronal heating 158
Coronal mass ejection 45,48, 188-189
Coronal void 70-71
Coronameter, K- 50-51
Criterion, Suydam 180
Current 51,110-111
    line 149-152
Current sheet 132-133,135, 138-140,146-147,152,181-182

Density 3,22-24,59-62,71,116

Differential cloud model 31
Differential emission measure 18,62
Differential rotation 44
Diffusion time, magnetic 10,131,167
Diffusivity, magnetic 9
Dimensions 3
Disappearance, sudden 4,7,38-45,124, 140,184,190
Displacement, footpoint 175
    Lagrangian 171
Doppler dimming 26
Doppler shift 30-31,68,109
Downflow 32
Dungey model 143
Dynamics 68

Eclipse 2,4,48
Emerging flux 42,44
Emission measure 64
    differential 18,62
Energy balance 63
Energy equation 9,130,158
Energy method 172-174
Environment of prominence 13,47-76
Equation, continuity 9
    energy 9,130,158
    Euler-Lagrange 174
    induction 9
    MHD 8-12
    motion 9
Equilibrium of prominence 13-14, 143-165
Equilibrium, hydrostatic 155-157
    thermal 155-164
Eruption of prominence 7-8,188-189
Eruptive instability 173-180
Euler-Lagrange equation 174
Evaporation 135-137
Evolution of prominence 19-22, 111-113

Feet of prominence 4,17,34-35,66,110, 119,125,127,129,140,165
Fibril structure 6,66,121,124
Filament, active-region (or plage) 1,3,4,7-8,15,17,20,44,96,116, 121,136-137
    winking 7,121,124
Filament channel 4,54-61,117,121,124
Filtergraph 29
Fine structure 4,17-19,21,65-66,70,

97,108-110,117,141,189
Flare, solar 7-8,39-40,45,48
Flows 31-34,41,43-45,75,110,168
Flux pile-up reconnection 132
Footpoint displacement 175
Force, self-pinching 150,152
Force-free field 11,152-154, 176,178
Formation of prominence 4,13, 115-141,183
Free-fall speed 10

Gaunt factor 63

H alpha profile 31
Hanle effect 78,81,83-85,87-88, 93-96,99,108
Heating 130
  coronal 158
Hedgerow prominence 23
Height, scale- 11,144
Helmet streamer 2,4,48-52,68, 72-75
Horizontal motions 34-37
HRTS spectrograph 30,34
Hydrostatic equilibrium 155-157

Impulsive bursty reconnection 132
Injection 135-137
Instability 12,38-46
Instability, ballooning 177, 180-181,188
  coalescence 132
  eruptive 178-180
  interchange mode 178
  kink 44,169,178
  MHD 130-132,168-170
  radiative 130,135,183-184
  Rayleigh-Taylor 170-172.178
  resistive 131,133,184-188
  sausage 168-169
  tearing mode 131,134,184-188
  thermal 130,135,183-184
Instruments 28-30,87-88
Interchange mode instability 178
Interface (or transition region), prominence corona 4,17-18,26, 30,34,36-37,48,53,59,61-66, 158

Internal structure of prominence 155-164
Inverse polarity 3,86,102,112,133, 138,141,149-152,164
Inversion (or neutral) line 3,71
Ionisation balance 60,65
Ionisation degree 22-24,116

K-corona 52
K-coronameter 50-51
Kink instability 44,169,178
Kippenhahn-Schluter model 3,34-35,45, 97,102,117,136,138,140-141,145, 149,155,164,179,181
Kuperus-Raadu model 3,34-35,45,102, 117,133,138,140,149,151-152, 163-164

Lagrangian displacement 171
Lambda meter 30
Lande factor 82
Lifetime of prominence 4,67
Line, neutral 3,54-57,91,93,106
Line current 149-152
Line tying 177-180,187-190
Linear pinch 168
Localised mode 169,180-181
Loop, post-flare 45-46
Loop prominence 45-46
Loops 17,34,125-126,128,134,141
Lyman continuum 25,27,60,159,164

Magnetic arch 50,52
Magnetic buoyancy 43
Magnetic configuration, three-dimensional 73,90
Magnetic diffusivity 9
Magnetic diffusion time
Magnetic field 13,24,36,52,54-55, 64-67,73,77-113,127
  force-free 11,152-154,176,178
  photospheric 90-93
  homogeneity 107-108
  inclination 99-102
  strength 93-99
Magnetic flux, cancelling 42,55,92, 117
  emerging 42,44
Magnetic polarity, inverse 3,86,102, 112,133,138,141,149-152,164
  inversion line 3,71

INDEX

Magnetic reconnection 34,35,
  66,74,92,132-135,139,
  149-150,185,189
  flux pile-up 132
  impulsive bursty 132
  Petschek 132,138
  Sweet-Parker 132
  unified 133
Magnetic Reynolds number 9,
  131,134,138-139,186
Magnetic roll 21,23
Magnetic shear 42,117,134-135
Magnetoacoustic wave 12
Magnetograph 77
Magnetohydrostatics 10-11,145
Marginal stability 171
Mass 47,68-70,75
Mass flux 28-38
Menzel model 143
Meridional circulation 23
MHD equations 8-12
MHD instability 130-132,168-170
MHD nonequilibrium 41
Microturbulence 26-29
Migration, prominence 55
Mode, ballooning 177,180-181,
  188
  localised 169,180-181
  Kippenhahn-Schluter 3,34-35,
   45,97,102,117,136,138,140-
   141,145,149,155,164,178,180
  Kuperus-Raadu 3,34-35,45,102,
   117,133,138,140,149,151-152,
   163-164
  normal 170-172
  resistive gravitational 184
  rippling 184
Model, differential cloud 31
  Dungey 143
  Menzel 143
  siphon 37,74-75,135
Motion, equation of 9
  prominence 89
Motions, helical (or twisting)
  43-45
  horizontal 34-37
  vertical 31-34,110
  vortex 41
MSDP spectrograph 30,32,37,55,89

Neutral (or polarity inversion)
  line 3,71
Nonequilibrium 174-176
NonLTE 24-26,82
Normal mode 170-172
Normal magnetic polarity 3,86,102,
  112,138,141,144-148,164

Observations 116-129
Optically thin radiation 17,161-162
Oscillations 37-38,90,165
Oscillator strength 63

Petschek reconnection 132,138
Pinch, linear 168
  self- 150,152
Pivot point 22,24,42,67,140
Plage filament 1,3-4,7-8,15,17,20,
  44,96,116,121,136-137
Plasma beta 10,74,138-139
Polar crown 4,18,92,94,96-97,102,
  106,138
Polarization degree 83-84,93
Polarity, inverse 3,86,102,112,133,
  138,141,149-152,164
  normal 3,86,102,112,138,141,144-
  148,164
Polarity inversion line 3,71
Pores 41,43
Post-flare loop 45-46
Prominence, active 1,62
  active-region (or plage) 1,3-4,
   7-8,15,17,20,44,96,116,121,
   136,137
  hedgerow 23
  loop 45-46
  quiescent 1
  winking 7,121,124
Prominence characteristics 15-22
Prominence environment 13,47-76
Prominence equilibrium 13-14,143-
  165
Prominence eruption 7-8,188-189
Prominence evolution 19-22,111-113
Prominence feet 4,17,34-35,66,110,
  119,125,127,129,140,165
Prominence fine structure 4,17-19,
  21,65-66,70,97,108-110,117,141,
  189
Prominence formation 4,13,115-141,
  183
Prominence lifetime 4,67

Prominence mass 47,68-70,75
Prominence migration 55
Prominence puzzles 12-14,76
Prominence types 1,143

Quiescent prominence 1

Radiation 10
  optically thin 17,161-162
Radiative instability 130,135,
  183-184
Radiative time-scale 130-131,140
Radiative transfer 25,56,68,158-
  160
Radio 58-61
Rayleigh-Taylor instability
  170-172,178
Reconnection 34,45,66,74,92,
  132-135,139,149-150,18_,
  189
  flux pile-up 132
  impulsive bursty 132
  Sweet-Parker 132
  Petschek 132,138
  unified 133
Resistive gravitational mode 183
Resistive instability 131,133,
  184-188
Resonance scattering 84
Roll, magnetic 21,23

Sausage instability 168-169
Scale height 11,144
Shear, magnetic 42,117,134-135
Sheet, current 132-133,135,138-
  140,146-147,152,181-182
Shock wave 12,132,138-140
Signal-to-noise ratio 88
Siphon model 37,74-75,135
Skylab 40-41,53,63
SOHO 46
Solar cycle 19-22,39,96,102,
  111-113
Solar flare 7-8,39-40,45,48
Solar Maximum Mission 19-20,30,
  34,68-69
Sound speed 10
Spectrograph 29-30
  HRTS 30,34
Speed, Alfven 131
  free-fall 10

sound 10
Spicules 136-137
Stability 14,125,167-191
  marginal 171
  thermal 17,19,183-184
Stokes parameter 79-82
Stokesmeter 78,85,88
Streamer, helmet 2,4,48-52,68,72-75
Structure, fibril 6,66,121,124
  fine 4,17-19,21,65-66,70,97,108-
    110,117,141,189
  internal 155-164
  three-dimensional 73,90,141,154,
    165
Sudden disappearance 4,7,38-45,183,
  190
Supergranulation 51,66-68,92,127-128
Surface Alfven wave 172
Surge 135
Suydam criterion 180
Synoptic map 102

Tearing mode instability 131,134,
  184-188
Tearing time-scale 131,167
Temperature 3,26-28,59
THEMIS 46
Thermal conduction 63,130
Thermal equilibrium 155-164
Thermal instability 130,135,183-184
Thermal stability 17,19,183-184
Threads 17-19,22,34
Three-dimensional structure 73,90,
  141,154,165
Time-scale, Alfven 10,131-132,167
  conduction 19,130
  magnetic diffusion 10,131,167
  tearing 131,167
Transition region 4,17-18,26,30,34,
  36-37,48,53,59,61-66,158
Turbulent velocity 26-28
Twisting motions 43-45

Unipolar region 21
Upflow 32,68,72

Velocity 21,28-38
  turbulent 26-28
Vertical motions 31-34,110
Void, coronal 70-71
Vortex motions 41

INDEX

Wave, Alfven 11,37,154,158
    magnetoacoustic 12
Waves 11-12
Winking filament 7,121,124
WKB method 180

X-rays 53,55,59

Zeeman effect 79-82,85,88,
    94-95,108